T0332441

Elementary Particles and the Early Universe

The birth of the Universe, and its subsequent evolution, is an exciting blend of Cosmology, Particle Physics, and Thermodynamics. This book, with its synoptic approach, provides an accessible introduction to these fascinating topics. It begins in Part I with an overview of cosmology and is followed by a discussion on the present understanding about the birth of the Universe, detailing the Planck Era, Inflation, and the Big Bang. It speculates the possibility of multiple Universes. Before moving on to explore the essentials of the Standard Model of Particle Physics in Part II, with particular stress on the electroweak force, the first example is that of acquisition of mass by gauge bosons via the Higgs mechanism. The book finishes in Part III with the thermal history of the Universe. This will also lead to understanding baryonic matter and baryogenesis as well as nucleosynthesis.

This book is suitable for those taking courses on particle physics, general relativity, and cosmology. Readers mathematically inclined, who wish to enhance their basic knowledge about the early Universe, will also find this book suitable to move up to the next level.

Elementary Particles and the Early Universe

A Synergy of Particle Physics and Cosmology in the Birth and Evolution of the Universe

Eitan Abraham and Andrés J. Kreiner

CRC Press is an imprint of the
Taylor & Francis Group, an informa business

First edition published 2023
by CRC Press
4 Park Square, Milton Park, Abingdon, Oxon, OX14 4RN

and by CRC Press
6000 Broken Sound Parkway NW, Suite 300, Boca Raton, FL 33487-2742

CRC Press is an imprint of Informa UK Limited

British Library Cataloguing-in-Publication Data
A catalogue record for this book is available from the British Library

ISBN: 9780367559960 (hbk)
ISBN: 9780367568412 (pbk)
ISBN: 9781003099581 (ebk)

DOI: 10.1201/9781003099581

Typeset in Times
by codeMantra

To our families,
Solange, Sebastian, Sofia, Nicolas
and
My mother Ingeborg, Ana María, Vicky, Javier

Contents

APPENDICES

Preface

Who is this book for?

For people who are interested in the Universe but have not had a formal education on the subject. It is also a supplementary book for students of mathematics, physics, and astrophysics. As such, this is not a textbook in the traditional sense because we have not included problems. Rather, we tried to create an informative monograph that explains ideas sometimes obscured in the literature.

Popular Science books play a very important role in the culture of a society. They provide potential readers a relatively easy access to otherwise complex subjects. In the light of this brief description, the present book is not at a popular level. Rather, it tries to transcend to the next level by including some mathematics, which is presented in a minimalist approach to avoid losing sight of the overall introductory nature of the book and focusing on ideas and concepts.

To take on two major pillars like Particle Physics and Cosmology and try to fit them in a short book makes the task of deciding what to leave out more challenging than what to leave in. Regarding particle physics, we chose to pay special attention to the creation of the first massive particles, namely the weak force gauge bosons, via the Higgs mechanism. Adapted for fermions, the same mechanism generates their mass. Simplifying, we see this as a "point of contact" between particle physics and the mass of the Universe. This constitutes a highlight of this book. We have left out important topics like quantum electrodynamics and quantum chromodynamics which are clearly beyond the scope of this book.[1] As far as the Universe is concerned, we concentrate on a particular time frame of its evolution, known as the Early Universe. This is quite a broad definition, so we define it between the origin, namely the Planck time $T_P = 10^{-45}\,\text{sec}$, and the first 375,000 years, when the release of the Cosmic Microwave Background took place.

We divided the book into three parts. Part I concentrates on Cosmology. Here we introduce some ideas that will be new to most of our readership. The interested reader will not have any problem in spending some time trying to think what the equations and concepts mean. In addition, in Part I we introduce the old Big Bang theory and inflation. In the 1980s, the theoretical discovery of a phenomenon called *inflation* redefined the Big Bang theory in a revolutionary fashion. It preceded the Big Bang starting at the Planck time $T_P = 10^{-45}$ seconds and finished at $t = 10^{-32}$ seconds, resulting in a volume expansion of 10^{53} of its original value. It gave rise to a different type of evolution of the Universe that opened the possibility of multiple Universes, a *multiverse,* which we discuss when we introduce inflationary cosmology. It is perhaps correct to say that for many readers this would be somewhat mesmerising. It is important to emphasise though that the physics in the Big Bang theory can be used to understand a number of phenomena without resorting to inflation.

In Part II we discuss the standard model of particle physics. We introduce spontaneous symmetry breaking and gauge symmetry, the two fundamental components of the Higgs mechanism. We will note that it is the vacuum expectation value of the Higgs field that provides the mass of particles and not the Higgs boson itself. We also give an introduction to the electroweak force which we complete between Appendices B and C.[2] This is another highlight of the book. In Part III, Particle Physics and Cosmology merge in the thermal history of the Universe, which we could call the third highlight. We discuss how the Grand Unified forces freeze out. In particular we discuss the electroweak force.

NOTES

1 Our next book, due in July 2023, will be at postgraduate level and will incorporate these topics.

2 Somewhat advanced Appendices suitable for the mathematically inclined readers.

Acknowledgements

One of us (EA) wishes to express his deep gratitude to Licenciada Fernanda Rodrigues-Gesualdi. Her unwavering support and encouragement during and after the lockdown phases of the pandemic was key to the completion of this book.

We wish to express our most sincere thanks to the following. Prof. Andrei Linde (Stanford University), for the trouble in locating the source of a figure used in his lecture notes. His advice on exploring the NASA website, which like for him, "made our stuff truly exciting". Prof. Ivo van Vulpen (Nikhef, University of Amsterdam) for authorising the use of pictures from the lecture notes by Marcel Merk, Wouter Hulsbergen, and himself. We also thank Ivo for his encouragement in pointing out that this book "is missing in the spectrum of available textbooks so far". Dr Stefan Luders (European Space Agency) for authorising us the use of a fascinating Planck image under the "ESA and the Planck Collaboration" copyright. His helpful attitude is truly noteworthy. Last but not least, our gratitude to the Argonne National Laboratory staff. To Dr Zein-Eddine Meziani, author of the beautiful figure showing the inner workings in the proton. This was possible thanks to the efficient action of Drs. Filip Kondev and Benjamin Kay which allowed us to get permission in a few hours. To all, once again, our sincere thanks.

Authors

Eitan Abraham is an Honorary Associate Professor of Physics at Heriot-Watt University in Edinburgh. Born in Israel and grown up in Argentina, he received his BSc from the University of Buenos Aires. The same year he became a postgraduate student at the University of Manchester and obtained his PhD in Quantum Optics. After postdoctoral work in Manchester and Edinburgh, he became an Assistant Professor at Heriot-Watt University. Initially, he did research in Theoretical Quantum and Nonlinear Optics. After a few years, his interests shifted to Josephson Junction circuits, High-Temperature Superconductivity, Magnetoelectricity, and cosmological applications of the Bohm–de Broglie formulation of Quantum Mechanics. He then joined the Institute of Biological Chemistry, Biophysics and Bioengineering at Heriot-Watt, where he investigated and proposed a model for the uptake of nanoparticles by cells. As a Physics Colloquia organiser between 2010 and 2012, he has invited eight Nobel Laureates and the Director-General of CERN, a month after the discovery of the Higgs boson in 2012. He has worked as a visiting scientist in European countries, the former USSR, and the USA. He has many years of experience in teaching nuclear and particle physics, quantum field theory, and general relativity to final year students. For 15 years he has been Director of Computational Physics. His main recreations include tennis, gym workouts, reading, and social life.

Andrés J. Kreiner is presently a Professor of Physics at the Universidad Nacional de San Martín (UNSAM), Head of the Accelerator Technology and Applications Department at the National Atomic Energy Commission (CNEA), and Senior Investigator at the

National Research Council (CONICET), Argentina. He grew up and lives in Argentina. He received his degree in Physics in 1973. Subsequently he studied Nuclear Physics at the Technical University in Munich receiving his PhD in 1978. He was a Research Associate at Brookhaven National Lab, USA, from 1980 to 1981 working at the Tandem Laboratory in Nuclear Structure problems. He was a visiting full professor and researcher at the Institute of Nuclear Physics at Orsay, and at the Centre of Nuclear Research at Strasbourg, France. He has been involved in basic and applied nuclear and accelerator physics R&D, and has taught continuously nuclear and modern physics, both at the degree and graduate level, from 1984 to 1994 at the Physics Department of the University of Buenos Aires. From then on he was at UNSAM where he organised the School of Science and Technology. Since 1994 he is involved in accelerator technology development for nuclear and medical purposes.

PART I

Cosmology

Introduction

The most incomprehensible aspect of the Universe is that it is comprehensible.
Albert Einstein

The magnitude of the intellectual feat achieved by humans in understanding physical phenomena defies comprehension. From particle physics to galaxies and all scales in between, the synergy between theory and experiment achieved unimaginable depths of knowledge. Encapsulated in this book is the attempt to convey the excitement that these observations generate in a unifying theme of Particle Physics and the Early Universe.

We begin Part I by presenting a very succinct historical development of Cosmology. Perhaps this can be considered the oldest science since it started with the interpretation of celestial bodies by early humanity. This is followed by a discussion on how we understand at the present time the birth of the Universe. From the Planck Era, inflation, and the Big Bang we have a picture of how it all started. In particular, we place emphasis on inflation, which is fundamental to everything that followed, and experimental observations, e.g., by the Planck telescope, consistent with it.[1] We discuss the fundamental importance of quantum fluctuations during inflation as the origin of galaxy formation. General Relativity is introduced via the Friedmann equations which encapsulate the Standard Model of Cosmology. When coupled with scalar field theory, the results are simply captivating. Inflation is followed by reheating, the more appropriate name for the Big Bang, which triggers the Radiation Era and thus the thermal history of the Universe.

Prior to discussing the thermal history in Part III, we give a synopsis of the Standard Model of Particle Physics in Part II. This is essential for understanding the freezing out of fundamental forces which are believed to have started under Grand Unification Theory (GUT) conditions after inflation ended. This will also lead to understanding, e.g., baryonic matter and baryogenesis as well as nucleosynthesis. As we shall discuss in Part III, the Last Scattering Surface was the start of the transparency of the Universe that released the Cosmic Microwave Background (CMB).

DOI: 10.1201/9781003099581-2

HISTORICAL PERSPECTIVE OF COSMOLOGY IN A NUTSHELL

An old philosophical question makes us wonder how do we know that we know what we know? The Universe offers an ideal platform for a debate. It has always prompted a fundamental question which mankind tried to answer throughout history: how did it all come about? Of course, there are many, many more questions regarding its size, composition, and so on, let alone our own existence. Sophisticated observational Cosmology in conjunction with General Relativity and Quantum Mechanics provides powerful tools to search for answers to these questions. Yet going back 13.7 billion years since the creation of the Universe, inevitably results in a mixture of scientific facts and conjectures. Hence, there is a boundary beyond which science gives way to speculative ideas, the validity of which can only be tested experimentally, perhaps, sometime in the future. It is then important to state from the outset that the ideas we present here about the origins of the Universe are not credo. They are a reflection of what we think we know and accept today in the Cosmology community.

The degree of success in understanding the physical Universe is perhaps best illustrated by a very brief historical perspective of Cosmology. With a few examples of the way of thinking in ancient times followed by highlights of scientific evolution and revolution, we will span some 4,000 years of history that led to Modern Cosmology.

Since the beginning of their existence as an intelligent species, human beings were always fascinated, if not mystified, by celestial bodies. Humans looked up to them as if they had god-like attributes. A pseudo-biological interpretation of the then known cosmos was that it formed from an egg or seed. A higher degree of sophistication was provided by mythology. It played a key role in the Babylonian version of Genesis and goes back to 1450 BCE. According to this version the sea is identified with disorder out of which deities emerge representing the sky, the sea, and so on. The story goes on with a fight in which the goddess of the sea is killed and out of her body the earth was created. In this vein, China also had its own brand of mythology arguing that a giant emerged from breaking an egg whose lighter parts formed the heavens and the heavier ones the earth. It is quite remarkable that there is a similarity in the mythologies from different parts of the world. A particular highlight is the common theme of order emerging out of chaos which, taken at face value, conceptually connects with examples in nonlinear dynamics and statistical mechanics.

Major progress came from the Greeks, and it can be said that Western Science has its roots in Greece. Despite the mythological component of their culture, they established basic principles for scientific enquiry. One of them was the identification of cause and effect. They also realised that observed phenomena could be phrased in a mathematical or geometrical rather than an anthropomorphic framework. Quite extraordinarily, Cosmology began to emerge as a scientific discipline against the general backdrop of rational thinking led mainly by Thales (625–547 BCE). The word Cosmology finds its origin in *cosmos*, meaning the world as an ordered system. The opposite to cosmos is the Greek word *chaos*. The advent of mathematical reasoning and learning about the physical world using logic and reason signalled the beginning of the scientific era. Aristotle (384–322 BCE) was one of the greatest thinkers whose influence lasted centuries. The crumbling of his geocentric cosmological thinking was started by Copernicus (1473–1543) with his heliocentric theory. It is quite extraordinary that he wanted to derive a single universal theory that treated everything on the same footing, an idea brought back to life by Einstein 400 years later. The final blow to Aristotelian ideas came from Kepler (1546–1601) who through painstaking observational astronomy established laws resulting from elliptical, rather than circular, planetary motion.

A quantum jump in cosmological thinking came from Newton (1642–1727). In his famous *Principia* (1687) he explained that Kepler's elliptical motion was the natural outcome of a universal law of gravitation. Newton's ideas dominated scientific thinking until the beginning of the twentieth century. The mechanistic view of Physics pervaded at all levels, even the propagation of light was assumed to need an *aether* to sustain it, an idea discarded by the Michaelson-Morley experiment (1887). Whereas for Copernicus and Kepler the solar system was the limit of human reach, in the eighteenth and nineteenth centuries the idea of a major structure they called Milky Way was taking shape. While this was considered the ultimate limit of the Universe, strange spiral "nebulae" were discovered across the sky that were very similar to the Milky Way. These objects would become known as *galaxies*. Clearly a conflict of interpretations emerged as the idea of more than one galaxy seemed inadmissible. This great debate at the beginning of the twentieth century was ended by Hubble (1889–1953), who discovered that the Milky Way was only one among hundreds of billions of galaxies.

Hubble's discovery and Einstein's Relativity marked the birth of modern cosmology. Einstein (1879–1955) in his Special Theory radically transformed the established ideas of space and time. His General Theory of Relativity revolutionised the law of universal gravitation. Einstein's field equations formulated in tensorial form are strongly nonlinear with analytical solutions not only challenging but also difficult to find. Schwarzschild (1873–1916) provided the first exact solution which ultimately led to the theoretical prediction of the existence of *black holes*. The first great works on relativistic cosmology were proposed independently by Friedmann (1888–1925), Lemaitre (1894–1966), Robertson (1903–1961), and Walker (1909–2001). They found an exact solution that describes a homogeneous and isotropic Universe; their model is the pillar of the Standard Model of Modern Cosmology. Another fundamental solution to Einstein's field equations came from de Sitter (1872–1934). He modelled the Universe as spatially flat, neglecting ordinary matter with dynamics dominated by the *cosmological constant* equivalent to the vacuum energy. It is thought that this constant in Einstein's field equations is a possible explanation of dark energy in our Universe.

Then came a mind-blowing hypothesis which evolved into theory. Its embryonic stage was started by Lemaitre who first noted that an expanding Universe might be traced back in time to an originating single point. Gamow (1904–1968) was an advocate of this idea and further proposed that the Universe started from a singularity and "exploded" into an energy inferno of ultra-relativistic particles and radiation. This became known as the *Big Bang*, a named coined by the British astronomer Fred Hoyle to illustrate the idea to a radio audience. This was a gigantic leap in understanding the early Universe. It was confirmed by the experimental discovery in 1965 by Penzias (1933-) and Wilson (1936-) of the CMB, a relic of 2.7K blackbody radiation from the Big Bang that constitutes the biggest clue to the early history of the Universe. The observation of CMB lent a spectacular support to the Big Bang theory, although, as we will see, the Big Bang can be interpreted as a reheating following the sudden "adiabatic" expansion of the Universe caused by inflation.

NOTE

1 We expect spectacular revelations by the James Webb Space Telescope. They were not available before the printing of this book.

Cosmology concepts

1

Experimental Cosmology has made some spectacular advances in the last few years. Several space telescopes have sent back to earth extremely valuable pictures and data. The latest releases from NASA's James Webb Space Telescope (JWST) have produced an excitement throughout the planet and an appetite to know more about the beauty and origin of our Universe. The estimated lifetime of JWST is 20 years during which information will be streamed. An example of a picture streamed from JWST is shown in Figure 1.1. Over 20 years of observations, we might have paradigmatic surprises that will change our understanding of the Universe.

We kick off by what it was the dominant view for decades regarding the birth of the Universe. Accordingly, it was triggered by the Big Bang some 13.7 billion years ago originating from a singularity at infinite temperature. There is considerable consistency between this cataclysmic cosmic fire and the observable Universe. The Big Bang is not just a qualitative description of an event. Rather, it is a very precise model based on the assumption that the early Universe was a hot gas in thermal equilibrium, and while expanding was pulled back by gravity.

From the Big Bang theory, we can work out at a given time the rate of expansion of the Universe, its temperature, and the densities of matter and radiation. It also explains how the light chemical elements H, He, and Li and their isotopes formed. The heavier atoms were created in the interior of stars where the high temperatures made fusion of nuclei possible. The most abundant elements in the Universe are H and He and the precision of the Big Bang theory enables us to calculate their abundances.

Yet some important facts remained unexplained. For example, the original Big Bang model assumes, but it does not explain, the large-scale homogeneity and isotropy of the Universe. In fact, looking around the sky what we get is, on average, a uniform picture in all directions to high accuracy to one part in ten. Rather than originating at a singularity, a reformulation of the theory concludes that the Big Bang happened everywhere and uniformly in the existing Universe at that time. How did this pre-Big Bang Universe come into existence? What was it made of? This will be answered as we navigate Part I.

DOI: 10.1201/9781003099581-3

FIGURE 1.1 This landscape of "mountains" and "valleys" speckled with glittering stars is actually the edge of a nearby, young, star-forming region called NGC 3324 in the Carina Nebula. Captured in infrared light by NASA's new James Webb Space Telescope, this image reveals for the first time previously invisible areas of star birth. Called the Cosmic Cliffs, Webb's seemingly three-dimensional picture looks like craggy mountains on a moonlit evening. In reality, it is the edge of the giant, gaseous cavity within NGC 3324, and the tallest "peaks" in this image are about seven lightyears high (Courtesy: NASA, ESA).

The old Big Bang theory tells us that this cosmic fire expanded and resulted in the Universe as we know it. It does not explain why the expansion occurred. Rather it is a theory that describes what happened in the aftermath of the Big Bang but not how it was engendered in the first place. Nor does the theory explain the origin of matter. All these remarks led to a paradigmatic reformulation of the very early Universe, namely that there must have been a scenario preceding the Big Bang: it is called *inflation*, a period of accelerated volume expansion of the Universe by an estimate of $\left[10^{24}\right]^3$ between 10^{-43} and 10^{-35} seconds. To give some idea of what this means, it is as if the diameter of a proton went from 10^{-15} m to 10^6 km. But how did it get triggered and what was the "seed"?

The answer will come as we develop Part I. We still have not discussed why inflation is important, but we can unequivocally say is that it became one of the pillars of modern cosmology. In an extraordinary fashion it can explain, for example, the isotropy and homogeneity of the large-scale Universe. At the same time, in conjunction with quantum mechanics, it can explain the lumpy structure of the Universe and the origin of galaxies based on causal physics. Implementation of inflationary ideas in its full glory leads to an interplay between particle physics, field theory, and cosmology.

1.1 COSMOLOGY ESSENTIALS

In order to explain properly the cosmological phenomena beyond a popular description, we need to resort to a minimum of mathematical rigour. To do this we have to introduce some basic cosmological concepts presented in this section.

1.1.1 Hubble's law

We begin with Hubble's law. It states that the further away is a galaxy, the greater the speed it recedes, that is,

$$\dot{R} = H_0 R \tag{1.1}$$

where R is the distance to the galaxy, the dot indicates time derivative, hence $\dot{R}$ is the speed of recession, with

$$H_0 \approx 70 \text{ km s}^{-1} \text{ Mpc}^{-1} \cong 21.8 \text{ km s}^{-1} \times \left(10^6 \text{ light years}\right)^{-1}$$

being the present value of Hubble's constant with Mpc = 3.09×10^{19} km called *megaparsec*. H_0 is referred to as a constant because it does not depend on spatial dimensions but is time-dependent. Furthermore, Hubble's law tells us that, in a homogeneous and isotropic Universe, the relative speed between two objects A and B separated by a distance r_{AB} is given by

$$\dot{\mathbf{r}}_{AB} = H_0 \mathbf{r}_{AB}$$

regardless of the orientation of $\mathbf{r}_{AB}$. That is, in a homogeneous and isotropic Universe there is no preferred origin or direction.

A fundamental quantity in standard cosmology is a dimensionless scaling parameter $a(t)$ which is uniform throughout space albeit time dependent. It is used to express an actual physical coordinate distance from, e.g., our planet to some distant galaxy at time t as,

$$R(t) = a(t) r_0 \tag{1.2}$$

where r_0 is a time-independent distance measured in a reference frame that is *comoving* with the expansion (*Hubble flow*). This coordinate system where distances remain the same is called the *comoving reference frame* fixed to the Hubble flow. In this frame, given two geometrically defined points as

$$\boldsymbol{x} = \{x_1, x_2, x_3\}$$
$$\boldsymbol{y} = \{y_1, y_2, y_3\}$$

then

$$\boldsymbol{x} - \boldsymbol{y} = \{x_1 - y_1, x_2 - y_2, x_3 - y_3\}$$

remains constant, and so does the distance given by Euclidean geometry,

$$|\boldsymbol{x} - \boldsymbol{y}|^2 = (x_1 - y_1)^2 + (x_2 - y_2)^2 + (x_3 - y_3)^2$$

as illustrated in Figure 1.2.

It has to be pointed out that the comoving reference frame, by its very nature, is a preferred reference frame. Against the background of a Universe without a preferred origin and direction, this would seem to be inconsistent. Yet it appears natural in the construction of the spacetime with the required symmetry properties.

Due to the expansion of the Universe, the distance between any two points is *not* constant but changes as $a(t)$. The comoving coordinate is fixed and can be likened to our familiar Euclidean coordinates for a flat Universe. The physical distances in our Universe are irreversibly stretched like a chewing gum. This stretching is uniform throughout the Universe, but it changes with *time measured by a clock in the comoving reference frame*. By replacing Eq. (1.2) in Eq. (1.1) we can rewrite Hubble's law as follows:

$$H = \dot{a}(t) / a(t) \tag{1.3}$$

and hence at any epoch, H gives the rate of expansion of the Universe.

Of crucial importance are the Hubble time and the Hubble radius. The Hubble time is defined as

$$t_H \equiv H^{-1} = 13.6\,\mathrm{Gyr}$$

and the Hubble radius,

$$R_H \equiv c t_H = 1.28 \times 10^{26}\ \mathrm{m}$$

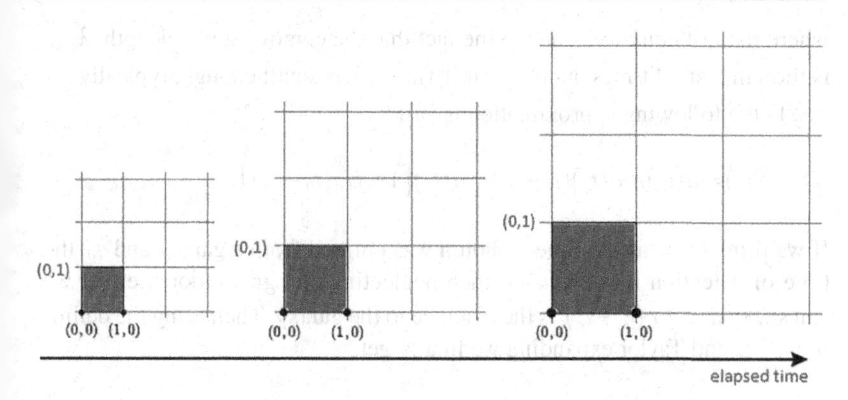

FIGURE 1.2 As time passes, the physical squares of this grid become larger, indicating the expansion of the Universe. The labelled points correspond to the comoving coordinate system showing that the distance between them remains constant. In this example, $x - y = \{x_1 - y_1, x_2 - y_2\}$ is constant when we substitute the values (0.0) and (1,0) at different times of the comoving coordinates. A comoving observer will see the squares remaining the same with time. However, the physical distance is the comoving distance times the scaling factor.

when we use the present value H_0. As we will see later, t_H is essentially the age of the Universe and R_H is the radius of the *Hubble sphere* which *defines the limit of the observable Universe*. Note that this is true for any observers, however far apart, who will have their own Hubble sphere and not necessarily in causal connection with each other.

1.1.2 Cosmological redshift

Wavelengths are subject to expansion through the scaling parameter and the relationship between physical and comoving wavelengths is, accordingly,

$$\lambda_{\text{phys}}(t) = a(t)\,\lambda_{\text{com}}$$

where λ_{com} remains constant. Then for a photon emitted at time t_e with wavelength λ_e and detected at time t_0 with wavelength λ_0, the *cosmological redshift* is defined as

$$z \equiv \frac{\lambda_0 - \lambda_e}{\lambda_e} = \frac{a(t_0)}{a(t_e)} - 1 \tag{1.4}$$

where the last equality is due to the fact that the comoving wavelength λ_{com} is the same at all times. Notice that if $(t_0 - t_e)$ is small enough, typically for $z < 1$, the following approximation is valid:

$$a(t_0) = a(t_e) + \dot{a}(t_e)(t_0 - t_e) = a(t_e)\left[1 + H_0(t_0 - t_e)\right]$$

If we think of t_e as the time a photon was emitted from a galaxy and t_0 the time of detection in our planet, then neglecting any gravitational effect we can say that $d = c(t_0 - t_e)$ is the distance to the galaxy. Then using the definition of z and Taylor expanding we finally get,

$$z \cong \frac{H_0}{c} d$$

which shows that redshift increases linearly with distance.

It is noteworthy that z, the fractional change in wavelength, coincides with the one predicted by the Doppler shift resulting from the relative velocity of the source (S) and detector (D), namely

$$z_{\text{Doppler}} = \frac{\lambda_D - \lambda_S}{\lambda_S} = \frac{v_S}{c}$$

whereas

$$z_{\text{Hubble}} \equiv z = \frac{H_0}{c} d = \frac{V_{\text{galaxy}}}{c}$$

where v_S is the speed of the source. z_{Hubble} represents an expansion effect (Hubble flow) whereby the photon travelling through spacetime experiences a redshift. Note the remarkable, perhaps not so surprising, identical expression to z_{Doppler}. Equivalently, since the photon momentum is given by $p = h/\lambda$, $p \propto a(t)^{-1}$ and the photon loses momentum with the expansion of the Universe. However, its speed is always the speed of light. On the other hand, for a massive particle with the momentum given by the de Broglie relationship identical to that of the photon, the loss of momentum means slowing down until the particle comes to rest in the comoving reference frame as the motion due to expansion never stops.

1.1.3 The peculiar velocity

As discussed, we express a physical distance in, say, the x-direction as

$$x_{\text{phys}}(t) = a(t)\, x_{\text{com}} \tag{1.5}$$

The rate of change of this physical distance defines physical velocity which naturally incorporates relative motion with respect to the comoving frame of reference. The velocity of this relative motion is called *peculiar*

$$v_{\text{phys}} = \frac{dx_{\text{phys}}}{dt} = a(t)\frac{dx_{\text{com}}}{dt} + \frac{da(t)}{dt}x_{\text{com}} = v_{\text{pec}} + Hx_{\text{phys}}$$

$$v_{\text{pec}} = a(t)\frac{dx_{\text{com}}}{dt}$$

and we used Eq. (1.3) to get the second term.

The physical velocity is made up of two components: the peculiar one and the Hubble flow. For a comoving observer when an object is in motion, it moves at a peculiar speed. If this latter is zero, the comoving observer never sees any motion. By contrast, the physical observer always sees motion resulting from the Hubble flow which never stops.

Horizons 2

2.1 THE HUBBLE HORIZON

If we can imagine a spherical surface that separates the observable from the unobservable, that surface constitutes a *horizon*. The *Hubble spherical surface* of radius R_H is a *horizon*. Suppose we wish to know the recession speed of a galaxy at a distance $R = \alpha \times 13.6\,\text{Gyr} \times c$, where α is a dimensionless number. Using today's value of the Hubble constant,

$$v_{\text{rec}} = H_0 R = \frac{1}{13.6\,\text{Gyr}} R = \alpha \times c$$

we have the following options depending on the value of α:

$$\alpha < 1 \Rightarrow v_{\text{rec}} < c; \qquad \alpha = 1 \Rightarrow v_{\text{rec}} = c; \qquad \alpha > 1 \Rightarrow v_{\text{rec}} > c$$

Hence galaxies within the Hubble sphere recede at speeds less than c and the emitted photons are detectable and there is causal connectivity. Those beyond recede at speeds greater than c and the emitted photons recede and are never detected (see Figure 2.1). As a result, causality is broken down. An observer might see regions beyond the present Hubble horizon if the Universe is decelerating and conversely (see discussion of Box 2.1).

Other fundamental definitions of horizons are the following.

2.2 PARTICLE HORIZON

Let us take a differential of comoving distance dr. Along the corresponding physical distance $a(t)dr$ it will take a time dt for light to propagate, that is,

$$cdt = a(t)dr$$

DOI: 10.1201/9781003099581-4

The comoving distance $r_{p-h}(t)$ between the initial time t_i when the Big Bang started and a final time t, e.g., the age of the Universe, is given by

$$r_{p-h}(t) = \int_{t_i}^{t} \frac{cdt'}{a(t')} \tag{2.1}$$

and the physical distance $d_{p-h}(t)$ is

$$d_{p-h}(t) = a(t) r_{p-h}(t) \tag{2.2}$$

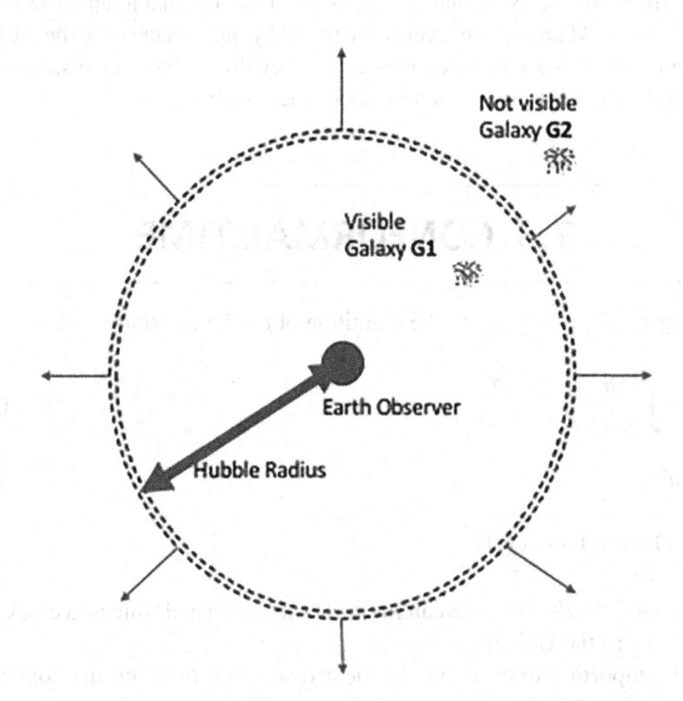

FIGURE 2.1 The Hubble surface seen as a sphere of radius R_H, the Hubble radius. Objects beyond this surface at distances R > R_H are receding from earth at super-luminal speeds and cannot be seen by observers on earth. Beyond R_H space is expanding faster than light and photons cannot reach an observer even though it is travelling towards it at the speed of light. According to Hubble's Law, the further is an object away, the faster it recedes from earth. Emission of photons inside the Hubble sphere is visible and hence objects inside it are causally connected. Objects outside cannot be seen but if R_H grows, it can catch up and previously unseen objects become visible.

Equation (2.1) defines the *comoving particle horizon*, or simply, *comoving horizon*. This is the comoving distance that light has travelled since the beginning of the Universe up to a time t. Equation (2.2) is the corresponding completed distance at time t. Thus $r_{p-h}(t)$ represents the comoving observable Universe. If we take objects separated by $r > r_{p-h}(t)$ today, then these objects were never in causal contact: a light signal will not have completed this distance at time t. We can say that $r_{p-h}(t)$ is the greatest comoving distance from which an observer at time t will be able to receive signals travelling at the speed of light emitted at t_i. If t is taken as the age of the Universe, then $r_{p-h}(t)$ is the maximum comoving distance travelled by light since the birth of the Universe. Note that $r_{p-h}(t)$ is not static and that it increases as the Universe ages. At any epoch, events separated by more than twice the particle horizon cannot have a common cause, i.e., they do not have a causal contact and therefore cannot possibly know about each other.

2.3 CONFORMAL TIME

Dividing $r_{p-h}(t)$ by c leads to the definition of *conformal time*,

$$\tau = \int \frac{dt}{a(t)} \tag{2.3}$$

and hence,

$$\tau(t) - \tau(t_i) = r_{p-h}(t) / c$$

If we write $d\tau = dt / a(t)$ we can interpret the conformal time as a clock that slows down as the Universe expands.

It is important to establish the clear distinction between the comoving Hubble radius $\mathcal{R}_H(t) = R_H / a(t)$ and $r_{p-h}(t)$. The comoving Hubble radius $\mathcal{R}_H$ is an upper limit for whether particles are causally connected or not:

- if they are separated by some comoving distance $r > \mathcal{R}_H(t)$, then currently they cannot communicate;
- if particles are separated by a distance $\ell > r_{p-h}(t)$, they could have never communicated with one another;

- if for some comoving distance $r > R_H(t)$ and $r_{p-h}(t) \gg R_H(t)$ the particles cannot communicate now but it is possible that they were causally connected earlier on but stopped due to an accelerated expansion through a mechanism explained in Box 2.1.

2.4 EVENT HORIZON

From the definition of particle horizon, we can conclude that there are past events we cannot see now. In a similar fashion we can think of future events we will never be able to see and distant regions we will never be able to influence. The way to quantify this concept is as follows. Consider a distant comoving observer B expecting a signal at a final time t_f and another commoving observer A sending a signal at time t. The largest comoving distance from which B can receive signals emitted by A at t is given by

$$r_{e-h}(t) = \int_t^{t_f} \frac{cdt'}{a(t')}$$

It can receive signals emitted at a time t' such that $t < t' < t_f$. In terms of the conformal time

$$\tau_f - \tau = r_{e-h}(t) / c$$

The distance $r_{e-h}(t)$ is the *comoving event horizon.*

We can illustrate these horizons with the help of Figures 2.2 and 2.3 by plotting a spacetime diagram. Given that $cdt = dr$ where dr is a physical distance, it can be recast as,

$$\frac{dt}{dr} = \pm\frac{1}{c}$$

where the $\pm$ stands for light propagating forwards (+) or backwards (−). The above equation defines two straight lines of slopes $+1/c$ and $-1/c$ that are the limits of causal connection. In 2D spatial dimensions they form a cone of light as illustrated by Figure 2.2.

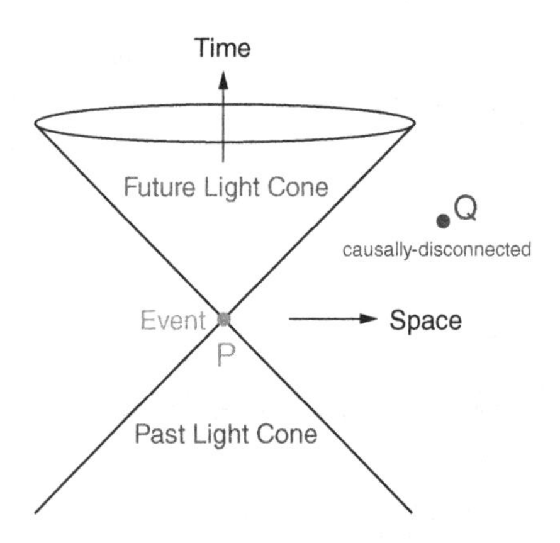

FIGURE 2.2 Time against 2D space. The set of all null geodesics (see text) passing through an event P form a light cone. All the events in the interior of or on the cone define a region of spacetime that is causally connected to P. Any event Q outside the cone cannot be reached from P as it would mean a speed greater than the speed of light. The separation between P and Q is said to be spacelike.

BOX 2.1: THE HUBBLE SURFACE AND VOLUME

Let us define R_H as the radius of a Hubble sphere. Its surface, the Hubble surface, bounds the volume and it is to be distinguished from the particle horizon which bounds the observable Universe (see below). We now succinctly explain how the causal mechanism depends on the properties of the Hubble sphere. As we show, the Hubble surface separates recession inside the sphere at a speed lower than c (subluminal) from the outside where recession occurs at a speed higher than c (superluminal).

The Hubble surface recedes at a radial speed,

$$\frac{dR_H}{dt} = c\frac{dH^{-1}}{dt} = c(1+q) \tag{2.4}$$

where

$$q = \frac{-\ddot{a}\,a}{\dot{a}^2} \tag{2.5}$$

When $q < -1$ the Hubble sphere contracts; when $q = -1$ it remains stationary, and when $q > -1$ it expands.

In a decelerating Universe (negative acceleration or $q > 0$) the Hubble surface recedes faster than galaxies and the Universe mass content increases. A galaxy at a distance $R > R_H$ recedes at a speed greater than c and when it is overtaken it ends up at a distance $R < R_H$ and then it recedes at a speed less than c: its superluminal recession becomes subluminal. The light emitted towards the observer by a galaxy outside the Hubble sphere recedes and thus it cannot be observed. At a later time, when it is overtaken, its recession becomes subluminal and can be observed.

In a linear expansion of the Universe ($q = 0$) the Hubble surface is stationary for a commoving observer and its mass content does not change.

In an accelerating Universe ($q < 0$) the galaxies recede faster than the Hubble surface and the sphere loses its mass content. Even if the acceleration is for a limited period, galaxies at distances $R < R_H$ are later at $R > R_H$ and their subluminal recession becomes in due course superluminal. Light emitted outside the Hubble sphere and travelling through space towards the observer does recede and can never enter the Hubble sphere and approach the observer: it becomes causally disconnected.

Each point on this diagram is called an *event* and each trajectory is a *worldline.* Massive particles travel along worldliness whose slopes at all points have absolute values greater than $1/c$. These are called *timelike geodesics.* Photons travel along worldliness called *null geodesics.*

In Figure 2.3 we represent conformal time against 1D comoving distance to illustrate the concept of particle and event horizons. Here the comoving observers do not move and therefore their worldliness are vertical lines emerging from each point in space. If we now recall that $cdt = a(t)dr$, we can write

$$c\frac{dt}{a(t)} = cd\tau = \pm dr$$

where the signs mean what we defined earlier. Furthermore, if we take $c = 1$, we get

$$d\tau = \pm dr$$

and graphically these are straight lines representing the propagation of light. They form an angle of 90° between them and 45° and 135° with the r-axis.

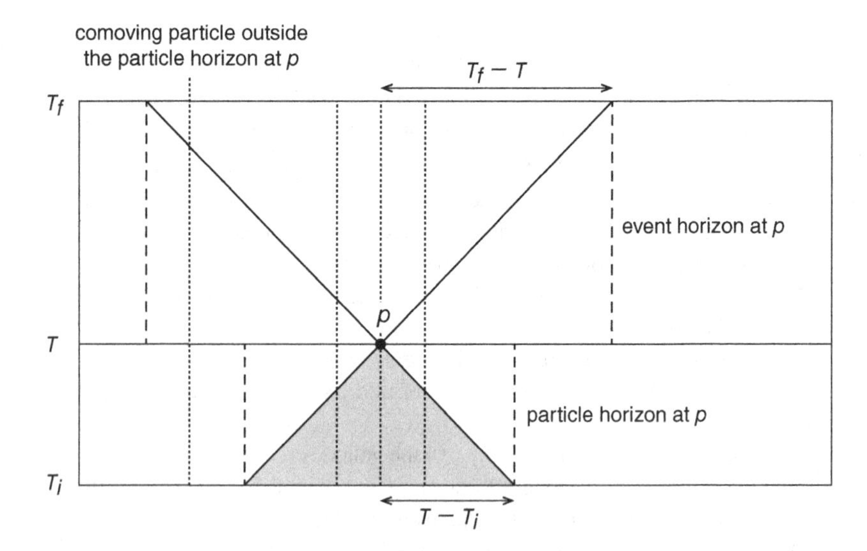

FIGURE 2.3 Spacetime diagram of conformal time against comoving distance to illustrate the horizons. Note that due to our choice of $c = 1$, the conformal time and the comoving distance are one and the same. This explains our labelling of the horizontal axis. As discussed in the text, the particle horizon is the maximum distance from which signals can be received by p. The event horizon is the maximum distance that p can send a signal to.

Again, the null geodesics going through an event p form a light cone. Given an initial conformal time τ_i and associating the time τ with the event p, the intersection of the lower cone surface with the τ_i horizontal line defines a segment twice the length of the particle horizon of p. Thus, all the comoving observers "under" the lower cone surface (the past), whose vertical worldliness intersect it, are causally connected to p.

If on the other hand we fix a time τ_f in the future, the intersection of the horizontal τ_f line with the light cone defines a segment twice the event horizon. As we discussed, this is the largest separation a future comoving observer can have from p to be causally connected with it. The intersection of the vertical worldliness with the upper light cone surface defines causal connections between p and the corresponding comoving observers.

3 The standard Big Bang theory

In this chapter we expand on our introduction to the birth of the Universe which was believed to start with a time singularity of infinite temperature. This was supposed to be the start of the Big Bang that became part of the standard Big Bang cosmology based on three foundations: the Cosmological Principle, Einstein's Theory of General Relativity, and a Classical Ideal Fluid Model of Matter. We will elaborate on these and sum up the successes and shortcomings of the theory.

3.1 FRIEDMANN EQUATIONS: EINSTEIN'S GENERAL RELATIVITY AND THE COSMOLOGICAL PRINCIPLE

At the core of the *Cosmological Principle*, it is assumed that over large distance scales the Universe is homogeneous and isotropic. By *homogeneity* we mean that every region of the Universe is essentially the same as any other region at the appropriate scales; there are no special locations. Clearly, on small scales, the Universe is inhomogeneous as it is full of planets, stars that are much denser than the interstellar medium, and galaxies that are much denser than the intergalactic medium. Yet, on scales larger than superclusters[1] and voids, the Universe is basically homogeneous. Average densities of matter and radiation within a sphere of radius of 100 Mpc are the same as the average density of any other sphere of the same size. Thus, the Universe has no edge or centre. The second assumption, *isotropy*, means that all directions are similar whichever way one looks. Once again, on small scales this symmetry breaks down. As with homogeneity, on scales larger than 100 Mpc,

DOI: 10.1201/9781003099581-5

the Universe is isotropic. There is also an implicit assumption that the laws of physics hold throughout the Universe. All these assumptions put together constitute the Cosmological Principle.

When the symmetries encapsulated in the Cosmological Principle are incorporated into Einstein's General Relativity, the **Friedmann equations** follow:

$$\ddot{a} = -\frac{4\pi}{3}G\left(\rho + \frac{3P}{c^2}\right)a \tag{3.1}$$

$$\left(\frac{\dot{a}}{a}\right)^2 + \frac{Kc^2}{a^2} = \frac{8\pi G}{3}\rho \tag{3.2}$$

where P is pressure, ρc^2 is the energy density made up of three components:

$$\rho = \rho_m + \rho_r + \rho_V \tag{3.3}$$

ρ_m : matter

ρ_r : radiation

$\rho_V = \dfrac{\Lambda}{8\pi G}$: vacuum

G is Newton's gravitational constant, Λ is the cosmological constant, $\dot{a}/a = H$ is the Hubble "constant", $a(t)$ is the dimensionless scaling parameter (defined by Eq. (1.2)), and K is the curvature of 3D space at constant time[2] (see Box 3.1). A quick inspection of Friedmann's equations shows an absence of space variables (homogeneity) and anything signifying a particular direction (isotropy). The only independent variable is time, and the unknown dependent variable is $a(t)$.

BOX 3.1: SPACE CURVATURES

Geometry of 3D space at constant time. From top to bottom: $K = 1, -1$, and 0, corresponding, respectively, to a closed, open, and flat Universe. $\Omega_0 \equiv \rho/\rho_c$ is the *closure parameter*: the ratio between energy density and critical energy density of the Universe: $\rho_c = 3H_0^2/8\pi G$, where H_0 is today's value of Hubble's constant. The numerical value of ρ_c expressed in different units is 1.9×10^{-29} g cm^{-3}, or, 1.1×10^{-5} protons cm^{-3}. The critical density of the Universe is the density borderline between eternal expansion and collapse. The latter occurs when $\Omega_0 > 1$.

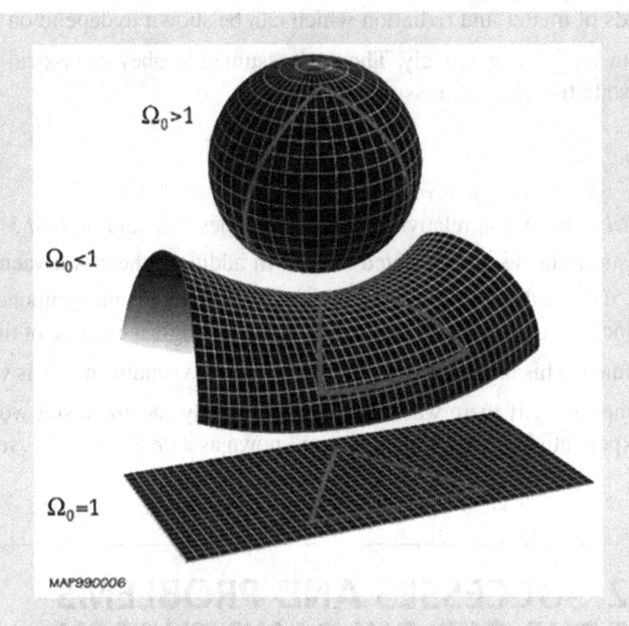

- $K = +1$ corresponds to a sphere, i.e., positively curved. Lines intersect. The sum of the angles of a triangle is greater than π. The volume of the Universe is finite, albeit it grows proportional to $a(t)^3$ only to collapse afterwards.
- $K = -1$ corresponds to a saddle, i.e., negatively curved. Lines intersect. The sum of the angles of a triangle is less than π. The space is infinite.
- $K = 0$ corresponds to a flat space, namely Euclidean that is the sum of the angles of a triangle is π. Parallel lines remain parallel but the distance between them increases due to the expansion of the Universe.

After some manipulation we get the continuity equation (conservation law):

$$\dot{\rho} \equiv \frac{d\rho}{dt} = -3\frac{\dot{a}}{ac^2}\left(\rho c^2 + P\right) \tag{3.4}$$

The Friedmann equations represent a huge simplification of Einstein's field theory, but they still are general relativity equations with homogeneity and isotropy incorporated. The source terms of these equations are cosmic uniform fluids of matter and radiation which can be shown to depend on $a(t)$ as $a(t)^{-3}$ and $a(t)^{-4}$ respectively. They are assumed to obey a classical gas equation of state that relates pressure and density:

$$P = w\rho c^2$$

where $w = 2/3$ for a non-relativistic gas of particles ρ_m and $w = 1/3$ for ultra-relativistic fluids as encapsulated in ρ_r. In addition, there is a vacuum contribution independent of time with $w = -1$. Although all components should in principle be present, depending on the cosmological era, one of them will be dominant. This latter applies to ρ_r and ρ_m. The vacuum energy is very small by comparison. If there was a Universe with only ρ_V the result would be an ever exponentially expanding Universe known as a *de Sitter Universe.*

3.2 SUCCESSES AND PROBLEMS OF THE OLD BIG BANG THEORY

The successes can be said to rest on three observational pillars. First, it explained the Hubble redshift-distance relationship we discussed earlier. Second, it predicted the existence and blackbody nature of the Cosmological Microwave Background (CMB). What is this? In a nutshell, when the Universe was about 380,000 years old, the temperature dropped to about 3,000 K. When the temperature was higher, the Universe consisted of an ionised plasma of protons, electrons, and photons. The mean free path of photons was much shorter than the Hubble radius due to scattering that rendered the Universe opaque. When the temperature dropped below 3,000 K, atoms formed and over a period of time the remaining photons crossed the Hubble radius, defined in Chapter 2, filled the Universe uniformly, and redshifted as the Universe expanded until today's 2.73 K. This radiation is known as CMB. The third observational success is the prediction of abundances of light elements and the observed ones. This tells us that from this period of formation or *nucleosynthesis* the Standard Big Bang cosmology works. This indicates that if something ought to be modified about it, we have to go to a period prior to nucleosynthesis.

The concept that it all started with a cataclysmic event from a singularity at infinite temperature is hard to accept but it has been a prevailing thought and justifiably so. After all, some of the predictions were successful as we explained above. Other issues include the size problem, the singularity problem, the degree of spatial flatness, the fluctuation problem, and the horizon problem. At this stage we will superficially concentrate on the last three because of their paramount significance.

3.3 THE HORIZON PROBLEM

The horizon puzzle can be best illustrated by the satellite measurements of the CMB blackbody radiation temperature (Figure 3.1).

What we notice in Figure 3.1 is an extraordinary uniformity which we get even looking in opposite directions. Although dark matter does not interact directly with radiation, it will interact through the gravitational force, leading to tiny "ripples" in the cosmic background radiation. This uniformity is puzzling as explained with the aid of Figure 3.2. Suppose for simplicity that the Universe is not expanding and take a point A which emits a photon at the very early Universe. This photon travels freely for 13.7 billion years to hit the North Pole of the earth. From a diametrically opposed and symmetric position, a photon is emitted from B which hits the South Pole. These photons could have never exchanged any information at the time of being released as they were separated by twice the age of the Universe. These photons are not causally connected and yet they must have communicated somehow because the temperature of the cosmic background radiation is almost the same in any direction and hence the photon energies.

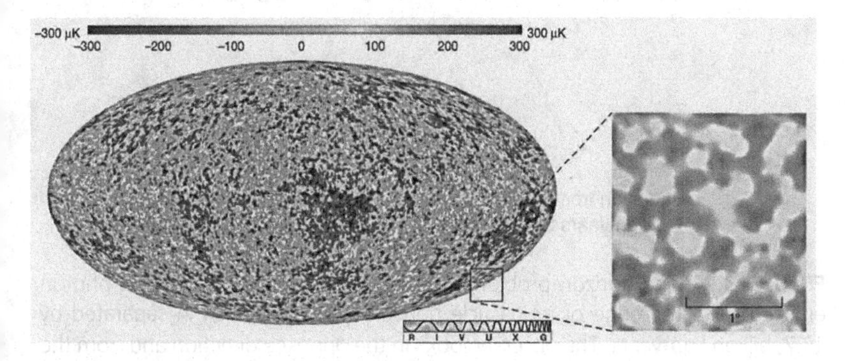

FIGURE 3.1 Cosmic background radiation temperature when the Universe was about 3×10^5 years old. Light grey and dark grey denote hot and cold variations of the 2.7 K temperature of the order of 1 part in 10^5 across the sky as shown by the horizontal scale. Only points separated by 2 geometric degrees (2°) or less are causally connected.

So why is the CMB so smooth? As we already pointed out, about 300,000 years after the Big Bang, the Universe cooled down to 3,000 K that led to the formation of hydrogen and helium atoms. Thomson scattering subsided and this marked the decoupling of photons from the primordial plasma and the Universe became transparent. As a result, photons were free to propagate and this we detect as CMB rescaled to 2.73 K. Whichever direction we point our detectors, the CMB radiation is virtually isotropic, with variations in temperature of one part in 10^5. In fact, it is established that the CMB is made up of some 10^4 disconnected patches of space, i.e., their comoving separations are greater than r_{p-h}. The fact that despite this they look so similar constitutes the *horizon problem*. The solution is provided by inflation.

In Figure 3.3 we show physical, as opposed to comoving, spacetime diagrams. Light cones compare schematically the standard Big Bang theory exhibiting the horizon problem and its inflationary solution. We take a present observer located at $x = 0$ with t_0 being the present time. Here the origin

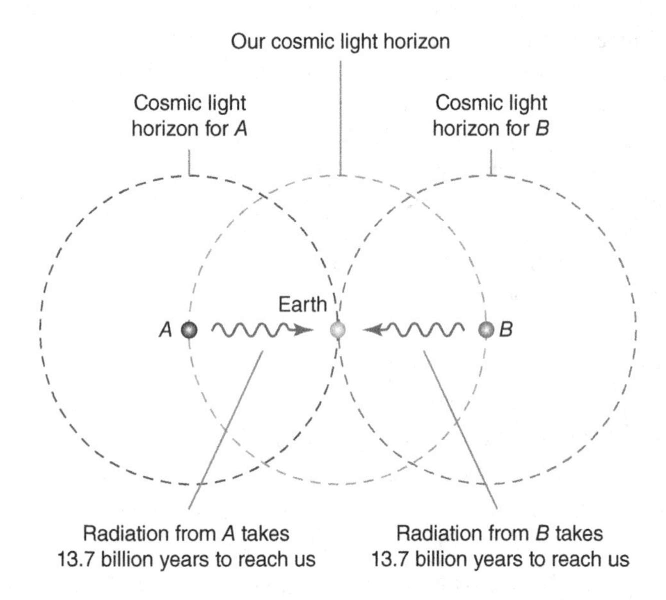

FIGURE 3.2 The horizon problem. The observer on earth receives a photon emitted from the edge of its particle horizon r_{p-h} emitted by A separated by 13.7 billion lightyears. The observer looks in the opposite direction and from the same distance r_{p-h} gets a photon from B. A and B are not causally connected as they are separated away some $2r_{p-h}$ from each other. They could not have communicated to produce almost identical blackbody emission spectra unless they did so at an earlier stage. This is the horizon problem which Standard Big Bang theory cannot explain. It was resolved by inflation.

is simply a generic point which could be anywhere. It is simply a reference point. The start of the CMB occurs at the recombination time $t = t_{rec}$, when atoms are formed, and photons are released never to interact again: they carry the information acquired then. The forward light cone $l_f(t)$ shown in Figure 3.3a comes from the Big Bang and the intersection occurs at $l_f(t_{rec})$ defining *a* physical event horizon between it and the origin. The segment so defined represents points that were in causal contact among them and our observer at earlier times than t_{rec}. The dark shaded region against the vertical axis shows this window of causality. Furthermore, we can draw another future light cone from $t = 0$ with its non-overlapping and hence disconnected event horizon: the yellow-shaded region indicates causality with the observer independent of the previous one. Of course, we can draw some intermediate light cone that can intersect the other two, but it will not establish causality between the yellow and blue regions. The important point to notice is that after a succession of intersecting light cones, we will find one that is disconnected from the first one. But how many are there? To answer this, we need to bring in the past light cone, $l_p(t)$, which at $t = t_{rec}$ defines *a* particle horizon for our observer. However, as we just discussed, within the particle horizon of our observer there are disjoint event horizons all connected with the observer but unrelated among themselves. Thus, in photons coming from these disjoint regions of causality with the observer, the temperatures of the photons will be unrelated. In fact, there are some 10^4 disconnected regions and yet the CMB shows a uniformity in temperature of one part in 10^5. This tells us that these regions must have had causal contact and this is what defines the horizon problem.

3.4 WHAT IS INFLATION?

Qualitatively described, inflation is a huge expansion of the Universe where the scale parameter $a(t)$ is proportional to $\exp(Ht)$. This happens in a very tiny fraction of a second, resulting in an expansion speed that exceeds the speed of light. This is not in contradiction with special relativity because there is no energy carried: it is a geometric expansion of spacetime. Under inflationary evolution, which starts at $t = t_i$, the forward light cone is hugely stretched. Inflation finishes at $t = t_R$ when the Big Bang time singularity is replaced by heating produced throughout the Universe producing a huge entropy, very high energy density, and a temperature of the order of 10^{15} GeV consistent with Grand Unified Theories (GUT). This cosmic fireball

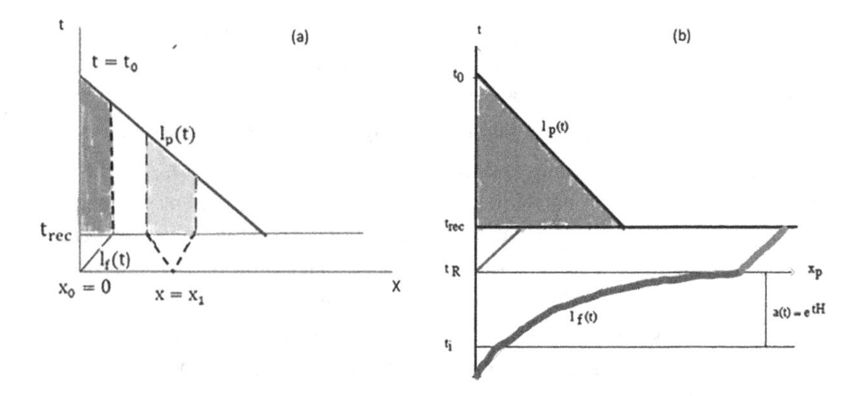

FIGURE 3.3 Physical spacetime light cones comparing schematically the standard Big Bang theory horizon problem and its inflationary solution. (a) The observer is located at $x = 0$ and t_0 is the present time. The CMB starts at the recombination time $t = t_{rec}$. The forward light cone $l_f(t)$ comes from the Big Bang and the intersection at $l_f(t_{rec})$ defines a physical event horizon. The points of this segment were in causal contact among themselves and the observer at $t < t_{rec}$. This is the dark-grey shaded region. We can draw another future similar light cone with its event horizon disconnected from the first one: the light-grey shaded region. The $l_p(t)$ light cone at $t = t_{rec}$ defines the particle horizon of our observer. (b) Under inflationary evolution which starts at $t = t_i$, the forward light cone is hugely stretched. Inflation finishes at $t = t_R$ and the Big Bang is replaced by a reheating (see text). The region covered is enormous enough so that the past light cone is contained within the forward light cone and all the points on the segment are causally connected.

over. The region covered is enormous enough so that the past light cone is contained within the forward light cone and all the points on the segment are causally connected. It is for this reason that t_R should be long enough to guarantee the connectivity that leads to a uniform CMB. It is estimated that the scaling parameter must increase by e^{64}: 64 *e*-foldings.[3]

Reheating is key as it is at this stage when the elementary particles of the Standard Model of Particle Physics are created.

3.5 THE FLATNESS PROBLEM

Now we concentrate on another puzzle: why is the Universe flat?

Suppose we consider a sphere of radius R. If we can change the radius of the sphere, we will notice that if the radius decreases the spherical nature becomes more evident. If on the other hand we increase the radius, the surface appears to be flatter. All this can be encapsulated in the definition of curvature, $\kappa \equiv R^{-1}$, namely the inverse of the radius. In the limit when $\kappa \to 0$, the spherical surface becomes a plane.

It is therefore intuitive to think that if inflation results in an accelerated expansion of the Universe, the effect should be one of flattening space as shown in Figure 3.4.

However, there is more than that and has to do with the density of the Universe. If we take the Friedmann equation (3.2) and write it in its equivalent form,

$$H^2 + \frac{Kc^2}{a^2} = \frac{8\pi G}{3}\rho$$

and suppose we take the present Hubble constant $H = H_0$ and solve for the density:

$$\rho = \frac{3H_0^2}{8\pi G} + K\frac{3c^2}{8\pi G a^2}$$

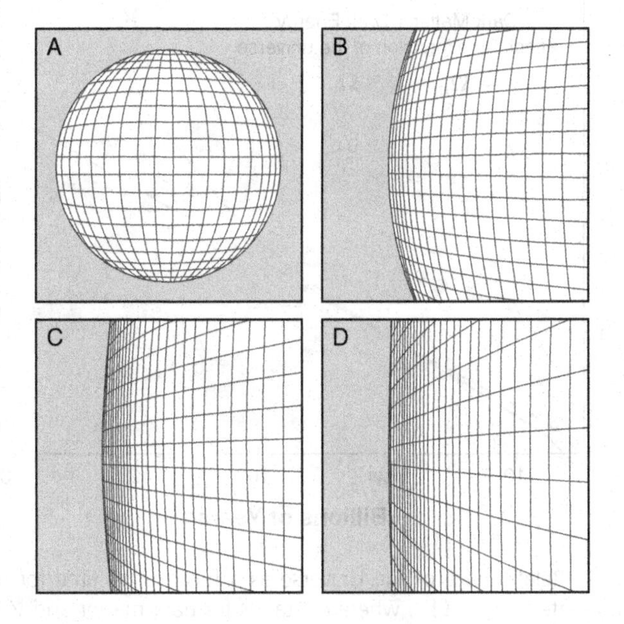

FIGURE 3.4 The expanding sphere illustrates the solution of the flatness problem by inflation (A-D). Due to inflation, we can now only see a tiny fraction of the Universe whose radius is over 13 Giga-lightyears. This is why the Universe looks homogeneous, isotropic, and flat.

From Box 3.1 we can identify the first term of the right-hand side as the critical density: $\rho_c = 3H_0^2 / 8\pi G$. Hence using the definition of the closure parameter $\Omega = \rho / \rho_c$ and the expression for comoving Hubble radius $\mathcal{R}_H = c / Ha$, we finally get

$$\Omega - 1 = K\mathcal{R}_H^2 \quad (3.5)$$

Some very interesting conclusions can be drawn from this expression. First, we trivially see that a flat Universe, $K = 0$, means that the density of the Universe is the critical one. If $K = +1$, then $\Omega > 1$, which results in a closed Universe evolving towards collapse Universe. If $K = -1$, then $\Omega < 1$ and we have an open Universe. These results are summed up in Figure 3.5.

In addition, the expression we obtained shows how inflation resolves the flatness problem. First, we notice that if inflation is an accelerated expansion

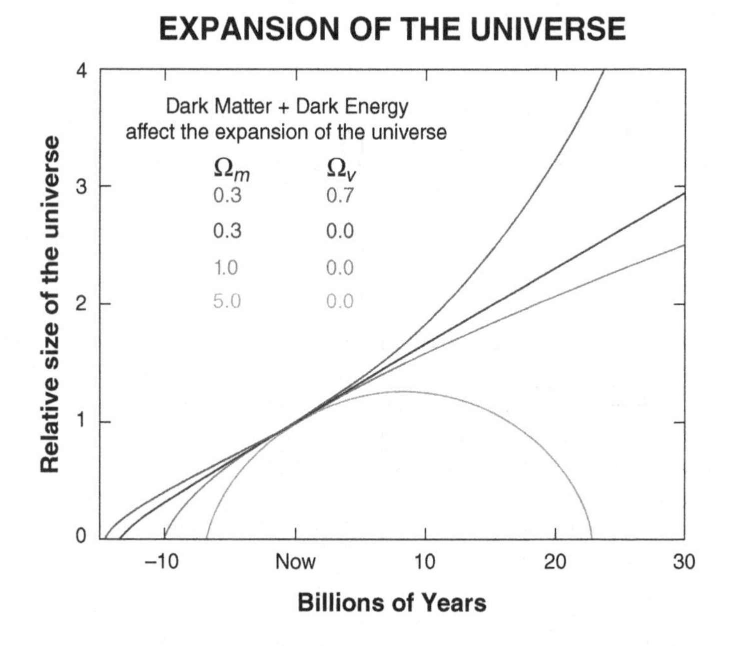

FIGURE 3.5 Relative size of the Universe as a function of time for different closure parameters $\Omega_m + \Omega_V$, where *m* stands for dark matter and *V* for dark energy. Third curve from the horizontal axis is Open ($K = -1$), second curve is Flat ($K = 0$), and first curve is Closed ($K = +1$). The top curve is an accelerated expansion assumed to originate from dark energy possibly coming from the vacuum energy. This is a de Sitter Universe (see text).

of the Universe, then from the Friedmann equation for the acceleration we must have,

$$\ddot{a} = -\frac{4\pi}{3} G\left(\rho + \frac{3P}{c^2}\right) a > 0$$

which means that

$$P < -\rho c^2 / 3$$

i.e., negative pressure or repulsive gravity. It also means $d\mathcal{R}_H / dt < 0$, namely a shrinking comoving Hubble radius since,

$$\frac{d\mathcal{R}_H}{dt} = \frac{d}{dt}\left(\frac{c}{aH}\right) = -c\ddot{a}\ \mathcal{R}_H^2 < 0$$

$$\Rightarrow \ddot{a} > 0$$

The implication of the shrinkage of $\mathcal{R}_H$ during inflation in Eq. (3.5) is that even if the Universe does not start flat, namely $\Omega \neq 1$, inflation drives it to flatness by making the right-hand side of (Eq. 3.5) vanishingly small. This solves the flatness problem.

NOTES

1 Superclusters form massive structures of galaxies, called filaments, supercluster complexes, walls, or sheets, that may span between several hundred million lightyears to 10 billion lightyears, covering more than 5% of the observable Universe. These are the largest structures known to date.
2 Note that the dimensions of *K* are 1/length^2.
3 *e*-folding is the time interval in which an exponentially growing quantity increases by a factor of *e*; it is the base-*e* analog of doubling time: $2 \times 2 \times 2 \times \cdots \times 2 = 2^N$ would be N two-foldings. In our case $\exp Ht$, $t = H^{-1}$, $\exp HH^{-1} = e \to 1\ e-$ folding. If $t = 64H^{-1}$ we get e^{64} i.e., 64 *e*-foldings.

Inflationary Cosmology I

4

Foundations

In this chapter we set out to answer some basic questions connected with cosmological *inflation*. What is it about? How is it produced? Why is it important? What are its consequences? To this end, we provide a basic theoretical framework for the inflationary scenario.

We begin by recalling that the Friedmann equations require an equation of state, namely one that relates pressure with energy density. To satisfy this requirement, a scalar field is introduced that will ultimately drive inflation.

4.1 THE SCALAR FIELD

In order to find solutions for the scaling parameter a, we need an equation of state that connects pressure with energy density. To do this, we bring in field-theoretic methods which, fundamentally, will provide a profound understanding of the origin of particles and radiation as we know them at present. It will also model the phase transitions leading to the freeze-out of fundamental forces in the Big Bang following inflation. In this section we will concentrate on a scalar field theory which will provide a model for inflation. This field is known as the *inflaton*.

The connection between the approach so far and a scalar field theory is that the energy-momentum tensor of a scalar field theory can be recast in a

DOI: 10.1201/9781003099581-6

way that looks like that of an ideal fluid. The inflaton or *scalar condensate* is thus characterised by an energy density,

$$\rho c^2 = \frac{1}{2}\dot{\phi}^2 + V(\phi) \tag{4.1}$$

where the first term represents the kinetic energy *density* and the second one the potential energy *density*. And for the pressure,

$$P = \frac{1}{2}\dot{\phi}^2 - V(\phi) \tag{4.2}$$

We now have to decide what kind of potential we adopt for the above equations. To do this we will adopt the most accepted inflation scenario known as *chaotic inflation*. According to it, the start of inflation does not require thermal equilibrium as the initial models, *old* and *new inflation*, required. The idea is that there is an initial distribution ϕ_i that is "chaotic", as ϕ_i can take on different values in different parts of the Universe. It can work with simple potentials like $V(\phi) \propto \phi^2$ isomorphic with the potential of a linear harmonic oscillator shown in Figure 4.1. It is not limited to polynomial potentials since chaotic inflation occurs in any theory where the potential can have a sufficiently flat region like that of Figure 4.2. In other words, there is a broad class of potentials that can work as long as they satisfy what is called *slow-roll* conditions plus a large initial value of the order of the Planck scale—concepts that will become apparent below. Before we use this potential note the following. If in Eqs. (4.1) and (4.2) we can neglect the kinetic energy, we obtain the equation of state,

$$P = -\rho c^2 \tag{4.3}$$

i.e., negative pressure. What does it mean? A gas in a balloon exerts positive pressure on the inside. If instead we tried to "suction" the surface of the balloon, that would be negative pressure. In other words, positive pressure is like compression, whereas negative pressure is tension. But what is the implication in our context? If we substitute Eq. (4.3) in Eq. (4.4), namely,

$$\ddot{a} = -\frac{4\pi}{3}G\left(\rho + \frac{3P}{c^2}\right)a \tag{4.4}$$

we get,

$$\ddot{a} = \frac{8\pi}{3}G\rho a \tag{4.5}$$

which is a positive acceleration indicating repulsive gravity which, in turn, causes expansion. Hence, we have the first ingredient of inflation, but we are not there yet.

Combining the above equations, we obtain the homogeneous-field Klein-Gordon equation with a friction term. This is equivalent to a differential equation of a harmonic oscillator with friction:

$$\ddot{\phi} + 3H\dot{\phi} + m^2\phi = 0 \tag{4.6}$$

Several regimes of motion are possible depending on the initial conditions. At the very top space-time foam of potential energy density, we have the regime of large spacetime fluctuations, i.e., the Planck scale as defined in Box 5.1 in Chapter 5. Below that, the light grey band is characterised by large inflaton quantum fluctuations. This leads to large jumps in ϕ leading to a process of *eternal self-reproduction* leading to a *multiverse,* i.e., multiple Universes. The darker grey band is about small quantum fluctuations. Finally, the bottom band leads to oscillations around the minimum leading to *reheating*, which we discuss in Chapter 5, that gives rise to pairs of elementary particle antiparticles and concomitant radiation. The Universe becomes hot.

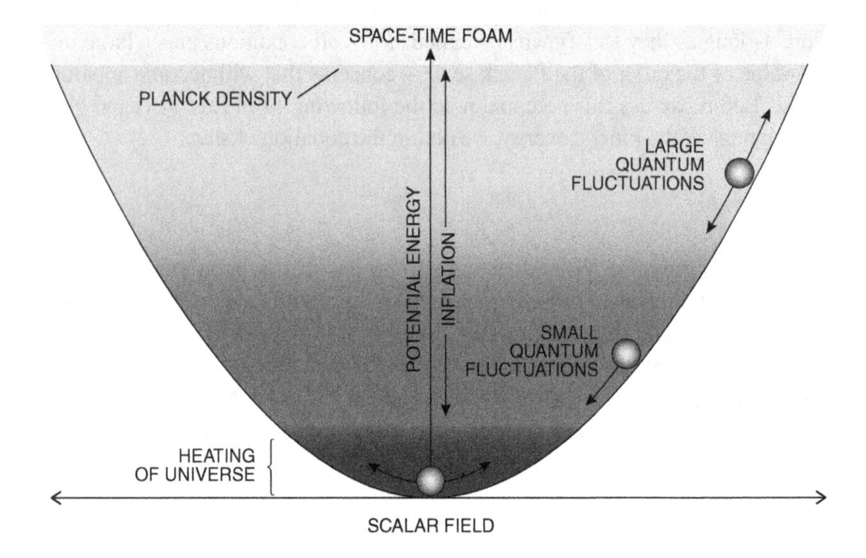

FIGURE 4.1 Harmonic oscillator inflaton potential $V(\phi) = \frac{1}{2}m^2\phi^2$ with an associated mass.

4.2 POTENTIAL ENERGY DENSITY PROPERTIES FOR INFLATION

The connection with General Relativity is established by inserting the energy density in Friedmann's equation. However, we assume conditions whereby we can neglect the kinetic energy density and only the potential one is entered.

Assuming a flat Universe, substitution of $V(\phi)$ in Eq. (4.7), namely,

$$\left(\frac{\dot{a}}{a}\right)^2 + \frac{Kc^2}{a^2} = \frac{8\pi G}{3}\rho \tag{4.7}$$

leads to

$$H^2 = \left(\frac{\dot{a}}{a}\right)^2 = \frac{m^2}{6}\phi^2 \tag{4.8}$$

In Eq. (4.6) we observe that in order for $d\phi / dt$ to be negligible we need a large friction, i.e., a large value of H. Then from Eq. (4.8) this implies a large value of $V(\phi)$, but what criterion do we apply to make this choice? Let us first note the following.

If the potential energy is large, then ϕ is large, and from Eq. (4.8), H is large and then the friction in Eq. (4.6) causes the field ϕ to move very slowly, rolling down the potential as a mass in mechanics. We can then assume that the potential remains constant, i.e., for a relatively long time so we can write,

$$H = \frac{\dot{a}}{a} = \frac{m\phi}{\sqrt{6}} \approx \text{constant} \tag{4.9}$$

The fact that ϕ is almost constant during inflation defines the slow-roll conditions. Integrating Eq. (4.9)

$$a(t) \sim e^{Ht} \tag{4.10}$$

which expresses the stage of inflation as an exponential expansion of the Universe.

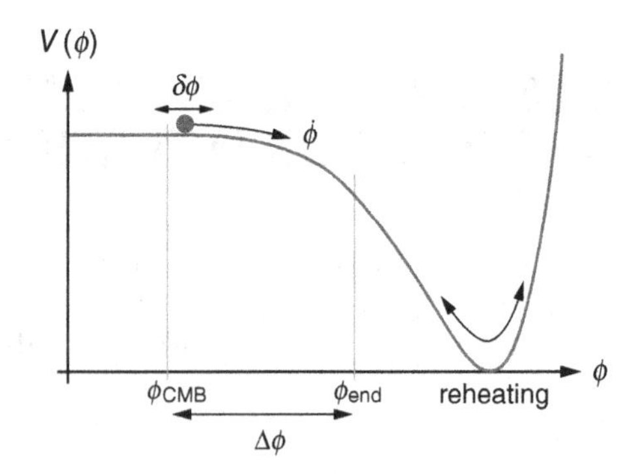

FIGURE 4.2 Non-polynomial potential energy density. Inflation occurs when the potential energy density dominates the kinetic energy density. Inflation stops at ϕ_{end} when the kinetic energy becomes comparable to the potential energy, i.e., $\frac{1}{2}\dot{\phi}^2 \sim V(\phi)$. CMB fluctuations are created by quantum fluctuations $\delta\phi$ some 60 e-folds before the end of inflation. Finally, at the minimum we have the reheating that gives rise to pairs of elementary particles-antiparticles and concomitant radiation. The Universe becomes hot.

Inflationary Cosmology II

5

Beginning and end

We still have questions to answer. How long does inflation go on for? What marks the end of this inflationary period? How do quantum fluctuations affect the inflationary scenario? Is there a possibility of a multiverse? We will give basic answers to these below and in Chapter 6, in the light of the hypotheses accepted presently.

5.1 GRAVITY

It follows from the discussion in Chapter 4 that the energy density of the scalar field remains approximately constant during inflation as ϕ is virtually constant and energy is accumulated as the volume expands. We will associate this energy as a kind of "matter" that creates negative gravitational energy so that the total energy is zero at all times. On the other hand, the volume of the Universe grows exponentially, since according to Eq. (5.1),

$$a(t) \sim e^{Ht} \tag{5.1}$$

and thus the volume is proportional to $a(t)^3 \sim \exp(3Ht)$. Then the energy of the scalar field grows at the same rate as the volume, namely $E \sim \exp(3Ht)$. After inflation ends, the scalar field decays to its minimum and all this

DOI: 10.1201/9781003099581-7

energy accumulated is transformed into the exponentially large energy/mass of particles populating the Universe. But why does it wait to decay before the release of its accumulated energy? The answer, not obvious at this stage, is that in order to release its energy it has to oscillate, and this effectively happens when it reached its minimum.

5.2 THE UNCERTAINTY PRINCIPLE: THE UNIVERSE IS BORN

How do we choose the initial inflaton ϕ_0 from where the process starts? Figure 5.1 tells us that the birth of the Universe is *conjectured* to be a space-time foam that lasted a Planck time $t_P = m_P^{-1}$ by taking $\hbar = c = 1$ (see Box 5.1). The foam was possibly formed by patches of energy and in order for each patch to be causally related it must have a Planck length $l_P = ct_P = t_P = m_P^{-1}$. Thus the energy density of the patch is $m_P l_P^{-3} = m_P^4$. In conclusion, the potential energy density at the start of inflation must be

$$\frac{1}{2}m^2\phi_0^2 \sim m_P^4 \sim 10^{94}\ \mathrm{g\,cm^{-3}} \tag{5.2}$$

Although the energy density at the start of inflation is extremely high, the total energy of the Universe is $\varepsilon = 0$.

Hence since energy is conserved, as the scalar field (i.e., matter) energy grows, it simultaneously creates a negative gravitational energy, thus

$$\varepsilon = +e^{3Ht}(\text{matter}) - e^{3Ht}(\text{gravity}) = 0 \tag{5.3}$$

It can be said that inflation starts with a small mass (or energy) $\Delta E \sim m_P \sim 10^{-5}$ g at the Planck time $t_P = m_P^{-1} = 10^{-43}$ s. This is clearly a far more acceptable start than the infinite energy and temperature of the old Bid Bang theory. But the question remains, where did m_P come from? It is at this point that we can go back to the uncertainty principle this time in Planckian units,

$$\Delta E \Delta t = m_P m_P^{-1} = 1 \tag{5.4}$$

Hence the emergence of the initial 10^{-5} g of matter is a result of the quantum uncertainty principle. Once we have this initial mass in the form of a scalar field, inflation is triggered, and the energy becomes exponentially large.

BOX 5.1: PLANCKIAN UNITS

Lenth: $l_P = \left(\frac{G\hbar}{c^3}\right)^{\frac{1}{2}} = 1.616 \times 10^{-33}$ cm

Time: $t_P = \frac{l_P}{c} = 5.391 \times 10^{-44}$ s

Mass: $M_P = \left(\frac{\hbar c}{G}\right)^{\frac{1}{2}} = 2.17 \times 10^{-5}$ g

Energy $\varepsilon_P = M_P c^2 = 1.2 \times 10^{19}$ GeV

Temperature: $T_P = \frac{M_P c^2}{k_B} = 1.5 \times 10^{32}$ K

Density: $\rho_P = \frac{M_P}{l_P^3} = 5.157 \times 10^{93}$ g cm^{-3}

In Planckian natural units with $\hbar = c = k_B = 1$

$l_P = t_P = M_P^{-1}$

$T_P = M_P \;\; \rho_P = M_P^4$

In Eq. (5.5), namely

$$\left(\frac{\dot{a}}{a}\right)^2 + \frac{Kc^2}{a^2} = \frac{8\pi G}{3}\rho \tag{5.5}$$

it is illustrative to consider the important *de Sitter Universe* for its implications in the interpretation of *dark energy.* This is a flat Universe where the only contribution to the energy density comes from the vacuum energy

$$\rho_V = \frac{\Lambda}{8\pi G} \tag{5.6}$$

where the *cosmological constant* Λ was introduced by Einstein to counteract gravitational collapse. ρ_V is constant throughout the Universe. In this case we get that the solution is

$$a(t) \propto \exp\left(\frac{\Lambda t}{3}\right)^{\frac{1}{2}} \tag{5.7}$$

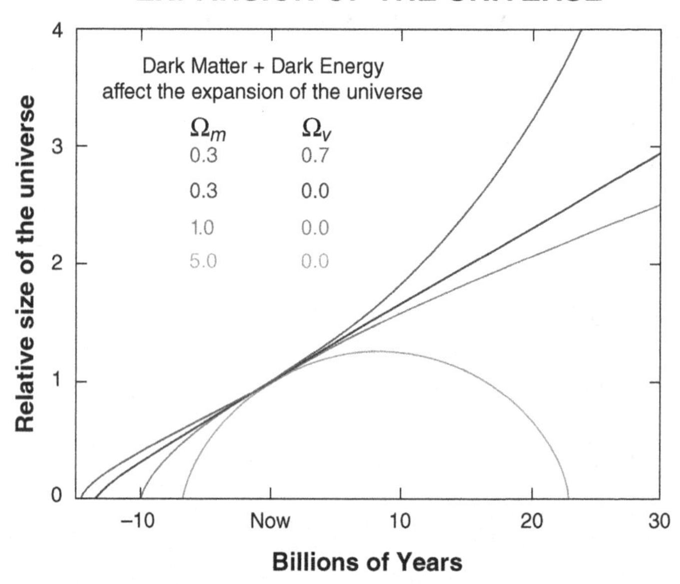

FIGURE 5.1 Relative size of the Universe as a function of time for different closure parameters $\Omega_m + \Omega_V$, where m stands for dark matter and V for dark energy. Third curve from the horizontal axis is Open ($K = -1$), green is Flat ($K = 0$), and bottom curve is Closed ($K = +1$). The top curve is an accelerated expansion assumed to originate from dark energy possibly coming from the vacuum energy. This is a de Sitter Universe (see text).

i.e., an accelerated ever-expanding Universe. Measurements confirm that our Universe is in accelerated expansion possibly caused by dark energy identified with the vacuum energy. The de Sitter model is not applicable to inflation as it never stops as it was shown in Figure 3.5 which we repeat here on Fig. 5.1.

Even so we describe the exponential time frame of inflation as a de Sitter stage.

The end of the inflationary period occurs at $t = 10^{-35}$ s when the reheating or "Big Bang" starts.[1]

NOTE

1 Note this is not the old Big Bang.

Inflationary Cosmology III 6

Quantum fluctuations and the origin of galaxies and multiple Universes

Fluctuations are key to life and ubiquitous in both physical and biological systems where atoms and molecules experience random motions. These kinds of fluctuations are of thermal origin, namely that the random motions occur because of a finite temperature. Another example is the fluctuations in the number of photons in a cavity in thermal equilibrium. However, there is a different type of fluctuation that is intrinsic and temperature independent and whose origin is quantum-mechanical. It is this latter that, defying the wildest imagination, is the origin of the large-scale structures of the Universe. How is this possible? In what follows we present the part of the answer within the framework of the inflationary scenario. We do this in order to stick to a chronological order of the evolution of the Universe. At the end of this chapter, we expand this answer to explain the tiny variations in temperature of the Cosmological Microwave Background (CMB).

DOI: 10.1201/9781003099581-8

6.1 QUANTUM FLUCTUATIONS

It is in fact the combination of inflation with quantum fluctuations that can create galaxies. Furthermore, inflationary fluctuations can create new exponentially large parts of the Universe. This process of *eternal inflation* can give rise to multiple Universes known as *multiverse.* Small quantum fluctuations of all physical fields exist everywhere, and we are interested in the quantum fluctuations of the scalar field. We can imagine these fluctuations as waves in the vacuum which appear and oscillate rapidly as in Figure 6.1a and then disappear. Inflation stretches them as it does to the Universe. When the wavelength of the fluctuations becomes sufficiently large, they stop moving and oscillating, but do not disappear. They look like frozen waves as we justify below with a simple mathematical model. When the expansion of the Universe continues, new quantum fluctuations become stretched, stop oscillating, and freeze on top of the previously frozen fluctuations as shown in Figure 6.2b and c. Why do fluctuations freeze? One can in principle think of it like this. As a result of the expansion the wavelength becomes larger than the Hubble radius, the *sub-horizon* part within the Hubble sphere and the *super-horizon* part outside the sphere become causally disconnected,[1] and the vibration stops.

As the fluctuation dynamics continues, a constructive superposition of these waves along the lines described in Figure 6.3b and 6.3c can cause a reversal of the slow-roll inflation. This pushes the scalar field up, causing a huge increase in the scalar field energy density as in the second from the top band of Figure 6.2.

We can show this pictorially for a 2D space in Figure 6.3a illustrating a landscape of eternal inflation. Huge peaks in the energy density trigger other inflationary scenarios which result in different Universes with possibly different physics laws and universal constants. In turn, inflationary processes repeat themselves in the corresponding new inflationary scenarios. In turn, they give rise to their own respective eternal inflations, thus creating further Universes as pictorially described in Figure 6.3b.

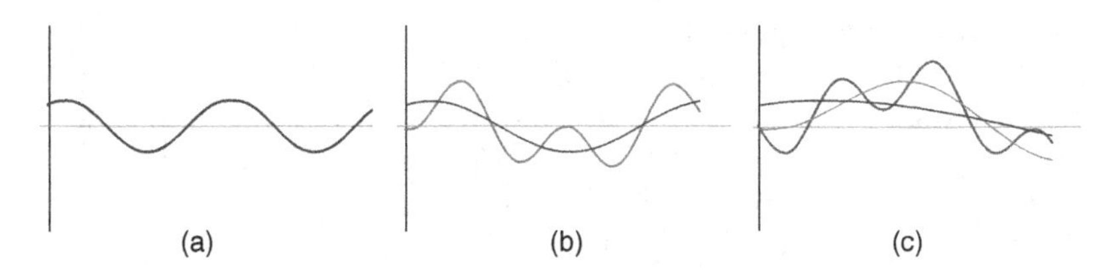

FIGURE 6.1 Quantum fluctuations in an inflationary Universe.

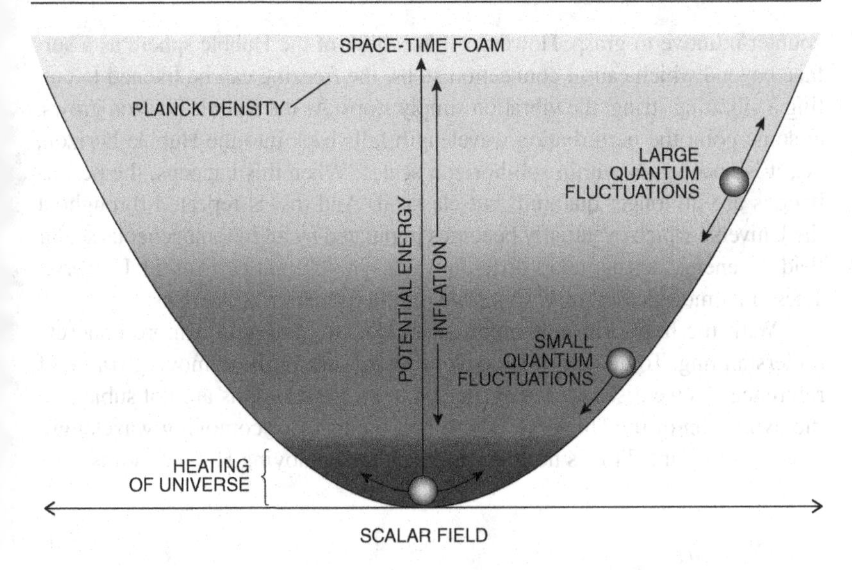

FIGURE 6.2 Energy bands, fluctuation dynamics and oscillations of the inflaton.

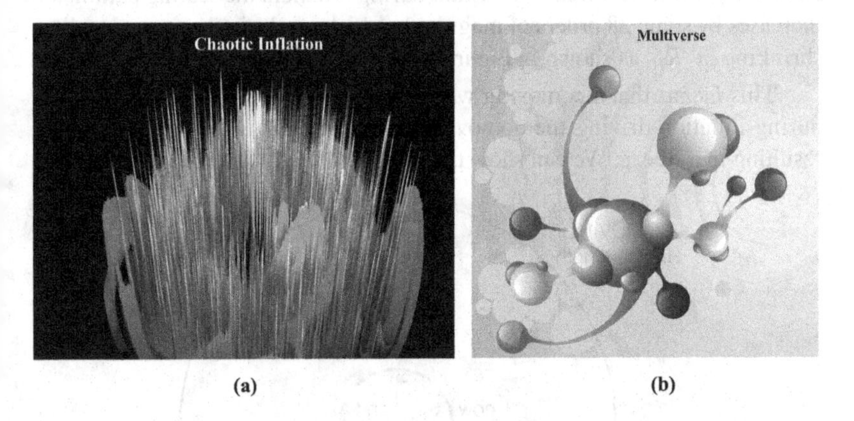

FIGURE 6.3 (a) Landscape of eternal chaotic inflation; (b) multiverse.

6.2 FREEZING OF FLUCTUATIONS

As we explained earlier, when during inflation the wavelength of the fluctuations is stretched beyond the limits of the Hubble radius, i.e., super-horizon scales, a freezing of the quantum fluctuations occurs. This freezing process is in principle counter-intuitive to grasp. However, if we think of the Hubble sphere as a surface beyond which causal connection stops, the freezing can be likened to cutting a vibrating string: the vibration simply stops. As the Hubble horizon grows, at some point the perturbation wavelength falls back into the Hubble horizon, i.e., it is once again within sub-horizon scales. When this happens, the perturbations are no longer quantum but classical. And this is repeated throughout the Universe which eventually becomes populated by an inhomogeneous scalar field. Its energy density takes different values in different parts of the Universe. These inhomogeneities are responsible for the formation of galaxies.

With the help of a mathematical model, we can gain a more concrete understanding. To do this, it is convenient to work in the comoving frame of reference. As we discussed in earlier sections, here lengths are not subject to the expansion of the Universe. Therefore, a particular comoving wavelength remains constant. This is not the case with the comoving Hubble radius,

$$\mathcal{R}_H = \frac{c}{aH} \tag{6.1}$$

Notice that while H remains constant during inflation, the scaling parameter increases by some 28 orders of magnitude which results in the corresponding shrinking of $\mathcal{R}_H$ as shown in Figure 6.4.

This means that a comoving wavelength $\lambda < \mathcal{R}_H$ can become $\lambda > \mathcal{R}_H$ during inflation driving the comoving perturbation to super-horizon scales resulting in a freeze. We can show this mathematically by considering a field

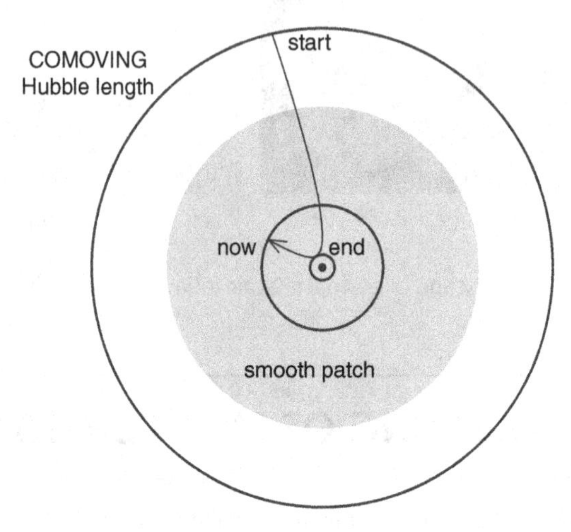

FIGURE 6.4 The comoving Hubble radius shrinks during inflation and grows afterwards.

φ representing the perturbation decoupled from the potential $V(\phi)$ which satisfies the Klein-Gordon equation with cosmological "friction". In physical coordinates it reads,

$$\frac{\partial^2 \varphi}{\partial t^2} + 3H\frac{\partial \varphi}{\partial t} - \nabla^2 \varphi = 0 \tag{6.2}$$

where we introduced the Laplacian to include spatial inhomogeneities. To move to comoving coordinates we convert to the conformal time τ recalling that $d\tau = dt / a$ and thus $\partial\varphi / \partial t = (\partial\varphi / \partial\tau) \times (d\tau / dt) = \varphi' a^{-1}$ and so on, where the dash means derivative with respect to τ. Applying this to the above and noting that Hubble's constant in conformal time is given by $H = a'a^{-2}$ we get,

$$\varphi'' + 2\left(\frac{a'}{a}\right)\varphi' - a^2\nabla^2\varphi = 0 \tag{6.3}$$

Note, for example, that $a^2\,\partial^2\varphi / \partial x^2 = \partial^2\varphi / \partial\left(xa^{-1}\right)^2$ but xa^{-1} is a comoving coordinate and hence the Laplacian now refers to variables of the comoving frame. With ∇^2 now referring to comoving coordinates we get,

$$\varphi'' + 2\left(\frac{a'}{a}\right)\varphi' - \nabla^2\varphi = 0 \tag{6.4}$$

To obtain a solution we expand the field φ in Fourier components,

$$\varphi(\tau, \mathbf{r}) = \frac{1}{(2\pi)^{\frac{3}{2}}}\int d^3k\left[\varphi_k(\tau)e^{i\mathbf{k.r}} + \varphi_k^*(\tau)e^{-i\mathbf{k.r}}\right] \tag{6.5}$$

Note that $\mathbf{k}$ is a comoving wavevector and $\mathbf{r}$ a comoving position vector. The *proper* or physical wavevector is $\mathbf{k}_p = \mathbf{k}a^{-1}$. Notice that $\mathbf{k}$ is not a dynamical variable but merely a label, as it were, for the Fourier component. Substitution leads to

$$\varphi_k'' + 2\left(\frac{a'}{a}\right)\varphi_k' + k^2\varphi_k = 0 \tag{6.6}$$

and similarly for the complex conjugate. We then introduce a mode function $u_k \equiv a(\tau)\varphi_k$ and using it in Eq. (6.6) leads to

$$u_k'' + \left[k^2 - \frac{a''}{a}\right]u_k = 0 \tag{6.7}$$

Using the acceleration Friedmann equation

$$\ddot{a} = -\frac{4\pi}{3}G\left(\rho + \frac{3P}{c^2}\right)a \tag{6.8}$$

and the equation of state we found above,

$$P = w\rho c^2$$

with $w = 2/3$, it can be shown that $a''/a = 1/\mathcal{R}_H^2$, i.e., the inverse squared of the comoving Hubble radius. We can substitute in Eq. (6.7) to get

$$u_k'' + \left[\frac{4\pi^2}{\lambda^2} - \frac{1}{\mathcal{R}_H^2}\right]u_k = 0 \tag{6.9}$$

To illustrate oscillation and freezing we take two extreme limits: a) $\lambda \ll \mathcal{R}_H$ and b) $\lambda \gg \mathcal{R}_H$

a) $\lambda \ll \mathcal{R}_H$ is the short wavelength limit or sub-horizon scale. In this case we can make the approximation to the differential equation,

$$u_k'' + k^2 u_k = 0 \tag{6.10}$$

This is the equation of a harmonic oscillator with an oscillatory solution

$$u_k = A_k e^{-ik\tau} + B_k e^{ik\tau} \tag{6.11}$$

b) $\lambda \gg \mathcal{R}_H$ is the long wavelength limit or super-horizon scale. Here the approximation is

$$u_k'' - \frac{a''}{a}u_k = 0 \tag{6.12}$$

which has the trivial solution $u_k \propto a$. Recalling that $u_k = a\varphi_{\mathbf{k}}$ we get

$$\varphi_{\mathbf{k}} = \text{constant} \tag{6.13}$$

This is what we set out to show, namely the mode freezing whereby a field mode with wavelength longer than the horizon stops being dynamical: asymptotically becomes a non-zero constant.

The simple model we used is surprisingly illustrative of what happens in the sub-horizon and super-horizon scales. With a more comprehensive theory we can obtain Figure 6.5 which illustrates clearly the separation between the two regimes.

The classically formed landscape results in gravitational attraction. Smaller peaks are attracted to bigger ones coalescing and forming large-scale structures of galaxies.

6.3 REHEATING

During the inflationary period the inflaton field acquires a huge amount of energy and the magnitude of the expansion redshifts away any possibility of matter and radiation. The Universe ends up in a cold state. However, we know there is a cosmic fire we call Big Bang and somehow this huge inflaton energy must be converted into particles and radiation. The mechanism for this to happen is called *reheating*. According to it, the inflaton field decays into Standard Model particles, fermions, and bosons,[2] as depicted in Figure 6.6.

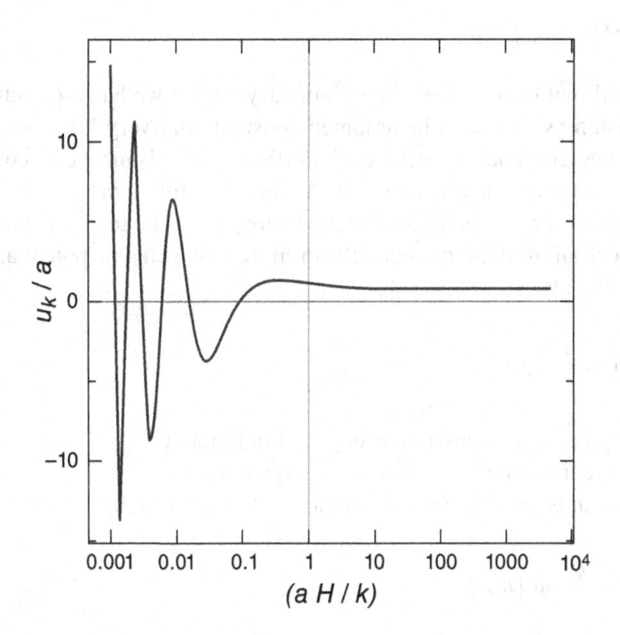

FIGURE 6.5 The normalised mode function showing oscillatory behaviour on sub-horizon scales $aH/k < 1$ and mode freezing on super-horizon scales $aH/k > 1$.

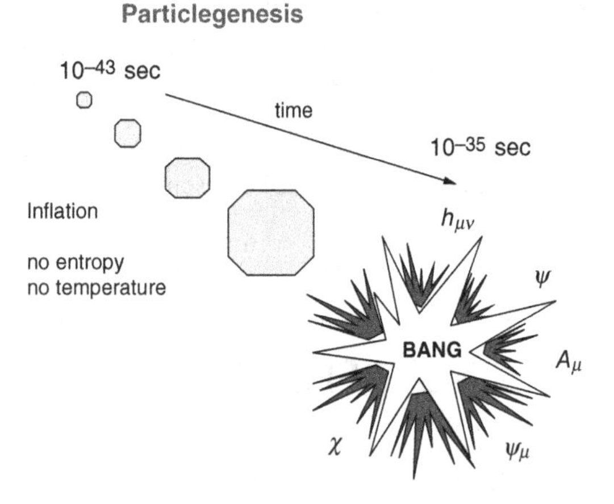

FIGURE 6.6 Artist impression of the transition from inflation to generation of Standard Model particles via reheating.

According to the Klein-Gordon equation we derived earlier, the evolution of the inflaton is given by

$$\ddot{\phi}+3H\dot{\phi}+m_\phi^2\,\phi=0$$

We argued that during most the inflationary period we had a de Sitter kind of expansion as H could be assumed constant, and very large, so that the acceleration term could be neglected. Furthermore, this meant we could consider the scalar field nearly constant. As this slow-roll approximation breaks down, H decreases, the field's kinetic energy is on the increase, and then it experiences an oscillation around the minimum of a chaotic potential we can approximate by

$$V(\phi)=\frac{1}{2}m_\phi^2\phi^2$$

The gently damped oscillation is depicted in Figure 6.7
which we reproduce more explicitly in Figure 6.8.
In fact, it can be readily shown that the solution is given by

$$\phi(t)\cong\frac{\phi_0}{t}\sin\left(m_\phi t\right)$$

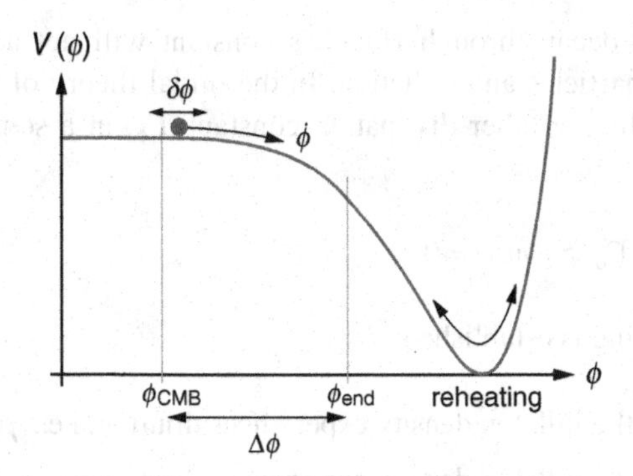

FIGURE 6.7 Preheating results in bursts of production of particles (n) as a function of time. These particles later interact and thermalize.

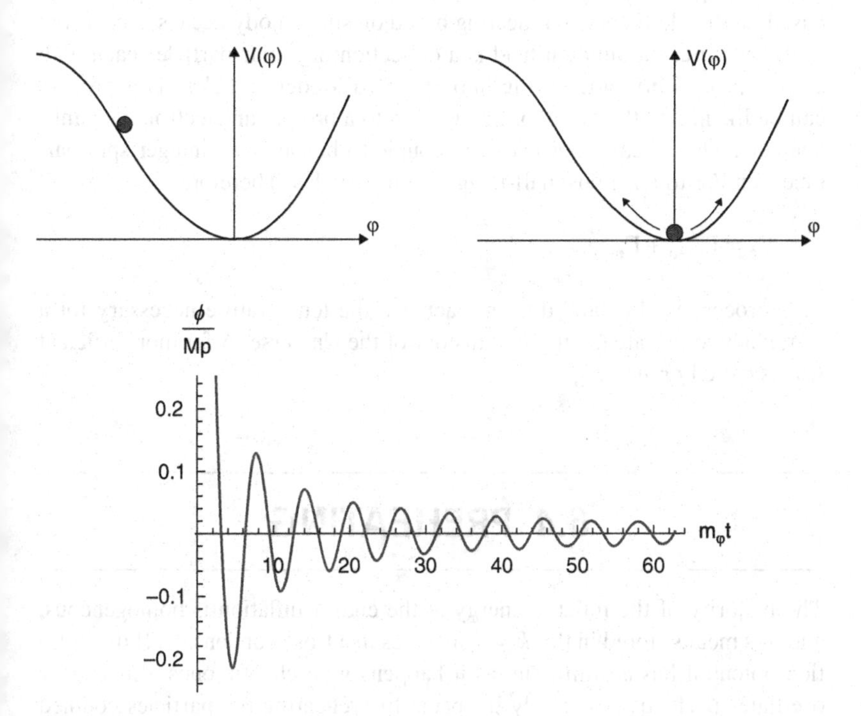

FIGURE 6.8 Oscillations of the inflaton around the with damping caused by the expansion, where Mp is the Planck mass defined in Box 5.1.

The expansion decay through Hubble's constant will not account for the production of particles and radiation. In the initial theory of reheating, this is done by adding another dissipation constant Γ_ϕ in a somewhat ad hoc fashion:

$$\ddot{\phi} + 3H\dot{\phi} + \Gamma_\phi \, \dot{\phi} + m^2\phi = 0$$

And the following is established:

- $H > \Gamma_\phi$ the inflaton density experiences dilution, i.e., $\rho_\phi \sim a^{-3}$
- $H \leq \Gamma_\phi$ the inflaton density experiences decay, i.e., $\rho_\phi \sim e^{-\Gamma t}$
- The temperature reached is the reheating temperature $T_r \sim \left(\Gamma_\phi M_P\right)^{\frac{1}{2}}$

The approach presented here is of course intuitive: the inflaton field must experience dissipation in the transition to Big Bang cosmology. This approach is based on the old theory of reheating based on single body decays. According to this picture, the inflaton field is a collection of scalar particles each with a finite probability of decaying into Standard Model particles. This process can be likened to the decay of the neutron to a proton, an electron, and antineutrino. These scalar particles can couple to bosons, i.e., integer spin particles, and/or to fermions, half-integer spin particles. Therefore,

$$\Gamma_\phi = \Gamma_{\phi \to b} + \Gamma_{\phi \to f}$$

This process is slow and does not achieve the temperature necessary for a symmetry to initiate the thermal history of the Universe. A far more efficient way is called *preheating*.

6.4 PREHEATING

The majority of the inflaton energy at the end of inflation is homogeneous, and this means stored in the $k = 0$ mode as in a Bose condensate. If the inflation potential has a minimum, as it happens with chaotic ones, this energy oscillates perfectly coherently in space. In preheating the particles coupled to the inflaton are resonantly amplified by parametric resonance as shown in Figure 6.9.

Since the occupation number of the inflaton for $k = 0$, i.e., its homogeneous part, is very large at the end of inflation, it can be treated as a classical.

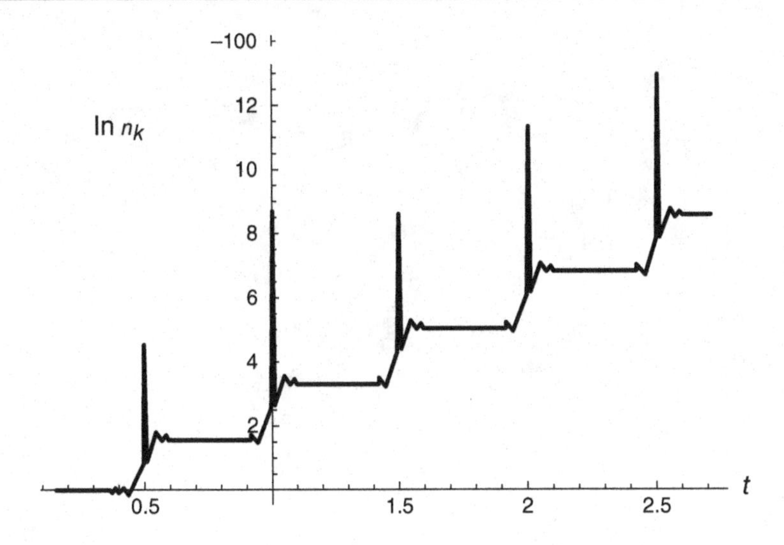

FIGURE 6.9 Preheating results in bursts of production of particles (n) as a function of time. These particles later interact and thermalize.

One can then think of it as a classical background force acting on the boson and fermion quantum fields. Since the inflaton is time dependent, this affects rapidly the effective masses of the bosons and fermions through parametric resonance. This causes a major departure from the old reheating theory we described above. This is what preheating is essentially about.

NOTES

1 See discussion on the Hubble horizon in Chapter 2.
2 See Chapter 7.

PART II

Particle physics

The Standard Model of Particle Physics I 7

Particles and fields

7.1 ELEMENTARY PARTICLES

The end of inflation marked the triggering of Big Bang cosmology. The assumption is that the oscillations of the inflaton have associated with them quanta, whose decay gave birth to what is known as the *Radiation Era*. Massless particles moving at the speed of light were the constituents of this radiation. There could have been all kinds of particles created at those extremes of temperature which we would not be able to observe today. The particles we will concentrate on are the ones we can observe that are the building blocks of nature. These, together with their antiparticles and the forces between them, form the foundation of the Standard Model of Particle Physics.

What is the Standard Model about? It tells us that *quarks* and *leptons* are *the* elementary particles of Nature upon which everything is built. It incorporates three of the fundamental forces: electromagnetic, strong, and weak. The symmetry aspects of the theory require these particles and the

DOI: 10.1201/9781003099581-10

force carriers of their interactions, known as *gauge bosons*, to be massless. Mass comes "externally" via the Higgs field, whose quanta are the Higgs bosons, and thus become part of the Standard Model. The corresponding antiparticles are also part of the model. Finally, the Standard Model incorporates the unification of the electromagnetic and weak interactions known as the *electroweak* theory.

To illustrate with some examples, protons and neutrons are made up of three quarks. Electrons are an example of leptons. Quarks come in six different flavours: *up*, *down*, *charm*, *strange*, *top*, and *bottom*. They all are spin-1/2 particles, i.e., fermions, and they have fractional charge given in Figure 7.1. Leptons are the *electron*, *mu*, and *tau* particles. They are too spin-1/2 fermions and have the same negative charge but, in addition, they have associated with them the corresponding *neutrinos* ν_e, ν_μ and ν_τ. These latter have no charge, are spin-1/2, and quasi-massless, but the small mass is very important as it is the cause of the so-called *neutrino oscillations*.[1] By some curious twist of Nature, these particles are classified in *generations* which differ in their mass but otherwise have basically the same properties. We can construct a periodic table shown in Figure 7.1.

The proton is constituted by *uud* quarks, whereas the neutron is made up of *ddu* quarks. The masses of the proton and neutron are around 1 GeV, which exceeds by far the sum of the masses of the *u* and *d* quarks (Figure 7.1).

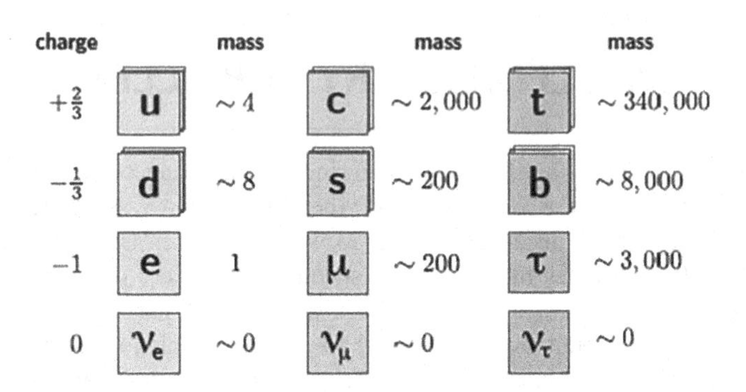

FIGURE 7.1 Three generations of quarks and leptons differing in their masses expressed here relative to the mass of the electron which is 0.511 MeV c^{-2} (using the special relativity expression $E = mc^2$ thus $m = E/c^2$). Quarks: Up (*u*), Down (*d*), Charm (*c*), Strange (*s*), Top (*t*), and Bottom (*b*). Leptons: Electron (e), Electron Neutrino (ν_e), Mu (μ), Neutrino *mu* (ν_μ), Tau (τ), Neutrino tau (ν_τ). All the charged particles have antiparticles with an opposite charge, e.g., the antielectron known as positron. Neutral particles, e.g., neutrinos and the photon, are their own antiparticles.

This hints at a very complex structure of the proton and neutron as shown in Figure 7.2. Thus, the *u*'s and *d*'s are *valence* quarks.[2] Both quarks and leptons have the corresponding *antiparticles* which have the same properties but opposite charge: when they come into contact, they annihilate each other. The annihilation results in the production of two photons of equal and opposite momenta. In turn, two photons, since they are their own antiparticles, can annihilate each other to produce, e.g., an electron-positron pair. This was the basis for thermal equilibrium in the early Universe. One more Standard Model particle of paramount importance that does not belong to this classification is the Higgs boson. This is a massive spin-0 particle, i.e., a scalar boson, and it is its own antiparticle.

Particles made up of three quarks are called *baryons* which also have their *anti-baryons*. The neutron and the proton are both baryons. Particles made of quark-antiquark pairs are called *mesons*. These pairs can in principle

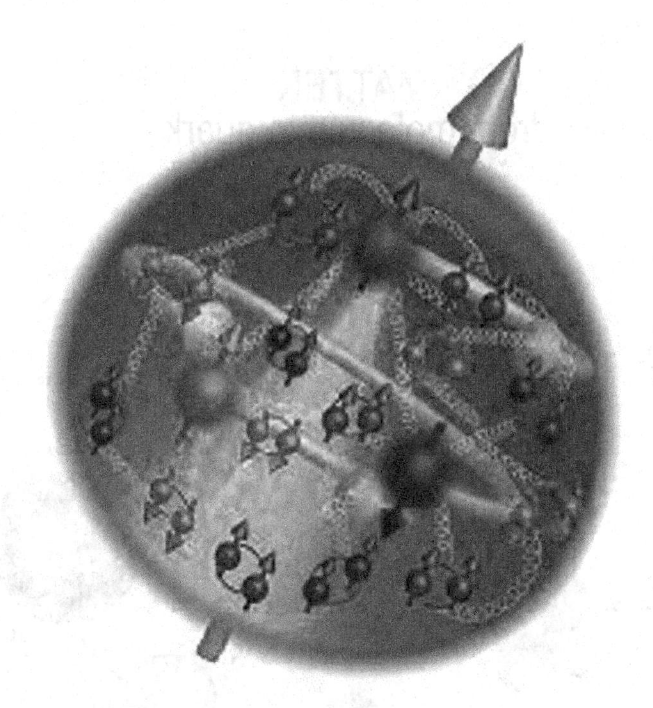

FIGURE 7.2 Rather than three main valence quarks (three larger spheres) connected by gluons, shown as springs, the proton's structure is much more complicated, with additional sea of quarks (small spheres) and gluons populating the proton's interior. Similar complex structure characterises the neutron (Courtesy: Argonne National Laboratory).

be made of any combination of quark-antiquark. Mesons and baryons are called *hadrons*. Finally, another important property of quarks is the *colour charge* which comes in *red*, *green*, and *blue*. They have their corresponding *anti-colours*. Quarks in baryons are always red-green-blue, i.e., white, and mesons are also colourless as the quark and antiquark have colour and correspondingly anti-colour charges. Table 7.1 shows different combinations expressed as three-component vectors of 1's and 0's. Although the leptons cannot be baryons, the combination of protons, neutrons, and electrons is referred to as *baryonic matter.* One of the greatest resolved mysteries in cosmology is the prevalence of matter over anti-matter known as *baryogenesis.*

Figure 7.3 shows the relative size of atoms, nuclei, and elementary particles.

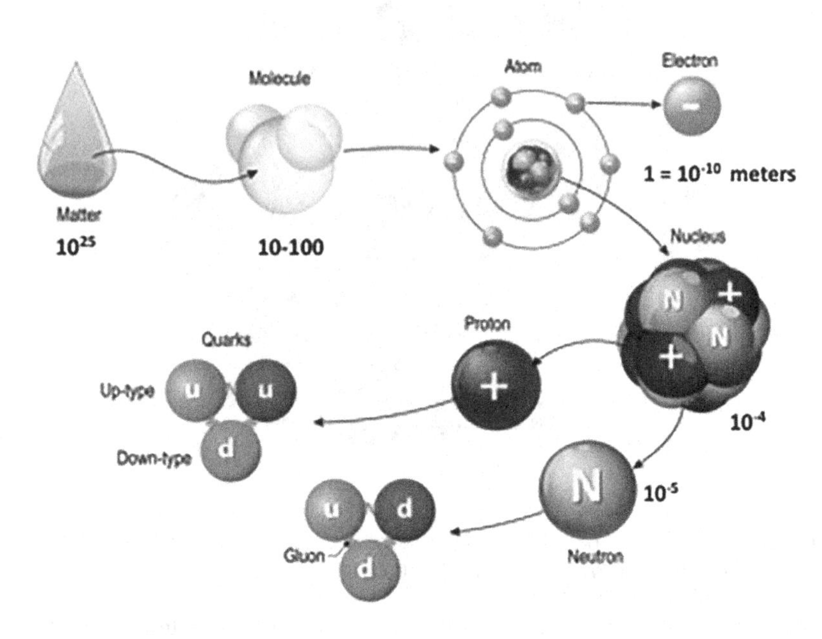

FIGURE 7.3 Relative sizes of particles and nuclei.

7.2 FUNDAMENTAL FORCES

Gravity, the fourth and most familiar of all forces, is missing in the Standard Model as fitting it into this framework proved intractable so far. The quantum theory of microscopic physics and general relativity appears to be incompatible to integrate them into the Standard Model. The fundamental forces are:

i ***Electromagnetic*:** between charged particles. The force is mediated by *photons*, particles of zero mass, thereby the range of the force is infinite. An example is illustrated in Figure 7.4.

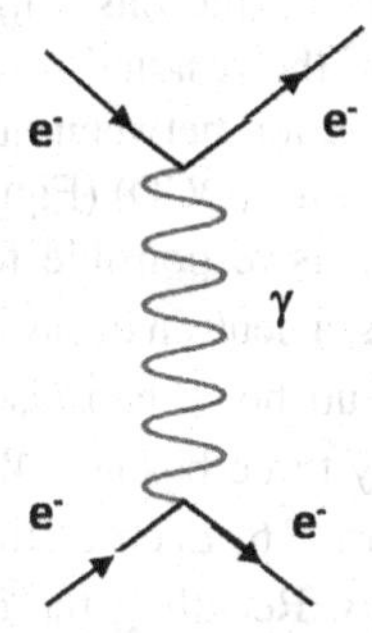

FIGURE 7.4 Feynman diagrammatic representation of the interaction between two electrons (full lines) via exchanging a photon. Time flows towards the right. Thus, the vertical wiggly line pointed by the arrow is an instantaneous interaction via a virtual photon, which is part and parcel of the interaction, and cannot be observed free.

TABLE 7.1

	RED	*GREEN*	*BLUE*	
Red	1	0	0	
Green	0	1	0	
Blue	0	0	1	
Cyan	0	1	1	= anti-red $\bar{r}$
Magenta	1	0	1	= anti-green $\bar{g}$
Yellow	1	1	0	= anti-blue b^{-}
White	1	1	1	
Black	0	0	0	

Baryons consist of $r + g + b = 1 + 1 + 1 =$ white (colourless)
Mesons consist of $r + \bar{r} = 100 + 011 = 111 =$ white (colourless)
(or $g + \bar{g}$ or $b + \bar{b}$).

ii **Strong nuclear force**: There is a *fundamental* strong nuclear force and a *residual* one. The former only occurs between coloured quarks and it is mediated by *gluons.*

These are chargeless and massless particles that carry both colour and anti-colour. There are eight of them. This force, therefore, only manifests itself within hadrons. As gluons are massless, the range of this force is infinite, but the force is attractive and increases with distance. The consequence of this is that there are no free quarks. However, after the Big Bang there was a quark-gluon plasma. In practice the range is of the order of ~10^{-17} m, i.e., the size of a hadron. A *residual strong force* exists between nucleons, namely proton–proton, neutron-neutron, and neutron-proton. This is mediated by three mesons called *pions* of charges −1, 0, and +1. The range of the residual force is about ~10^{-15} m. The field studying the interaction between quarks via gluons is called *Quantum Chromodynamics* (QCD) (Figure 7.5).

iii **Weak nuclear force**: It is responsible for radioactive decay, e.g., $n \rightarrow p + e^- + \overline{\nu_e}$, that is, a neutron decays into a proton, an electron, and an electron anti-neutrino. It can also change the flavour of a quark. It is mediated by three bosons W^+, W^-, and Z, as shown in the example of Figure 7.6. Each of them has spin 1 and hence known as *vector bosons.* Recalling the relativistic energy equivalent of mass, $E = m_0c^2$, we can express the masses of the W's as m_w = 80 GeV c^{-2}. They are oppositely charged $\pm e$ and are each other's antiparticles. The Z^0 boson is neutral with a mass m_Z = 92 GeV c^{-2}. The apparent violation of the conservation of energy by so

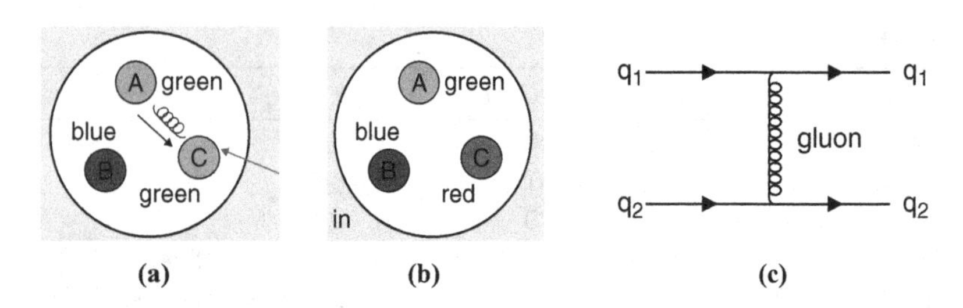

FIGURE 7.5 The strong force between confined quarks is mediated by gluons which interact with their colour charge. Gluons carry a colour and an anti-colour. (a) For illustration purposes we show a baryon with a colour imbalance as one of the green quarks should be red. (b) A emits an $r\overline{g}$, i.e., a red–anti-green gluon to C so that $g[\text{quark}] + r\overline{g}[\text{gluon}] \rightarrow r[\text{quark}]$. (c) Feynman diagram representing the interaction between quarks.

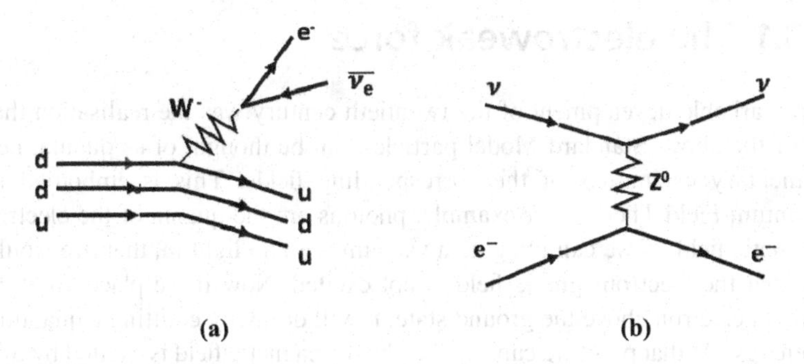

FIGURE 7.6 Examples of the weak force in action. (a) Neutron (ddu) decays into a proton (uud), an electron, and an anti-neutrino via the decay of the W^- boson. (b) Interaction of an electron with a neutrino via exchanging a Z^0 boson.

TABLE 7.2 Summary of Standard Model particles and the fundamental forces they are affected by

	WEAK	*EM*	*STRONG*
Quarks	√	√	√
Charged leptons	√	√	×
Neutral leptons	√	×	×

heavy gauge bosons is explained away by invoking the time-energy uncertainty principle. The range of the force is ~10^{-18} m, and it only affects particles spinning counter-clockwise while going away, and antiparticles spinning clockwise while going away.

iv **Gravity:** If we think in Newtonian terms, this is a force acting between all massive particles. It is clearly the most familiar, the weakest, and possibly the least understood. The force is only attractive, it has an infinite range, and string theorists advocate that, in the context of quantum gravity, it is mediated by a massless spin-2 *tensor boson* called *graviton.*

Table 7.2 sums up the forces quarks and leptons are affected by.

7.3 THE ELECTROWEAK FORCE AND THE HIGGS BOSON

To complete the description of the Standard Model, we have to bring in the *electroweak force* and the *Higgs boson.*

7.3.1 The electroweak force

A remarkable development of the twentieth century was the realisation that all of the above Standard Model particles can be thought of as quanta, i.e., elementary excitations of the corresponding fields. This is embodied in Quantum Field Theory. For example, photons are the quanta of the electromagnetic field. If we can imagine a vacuum with no light in, then we would say that the electromagnetic field is not excited. Now if we place an atom with an electron above the ground state, it will decay by emitting a quantum of energy. At that point we can say the electromagnetic field is excited by this quantum we call photon. In a similar fashion electrons and quarks are the quanta of their own fields. Table 7.3 gives examples of fields and their quanta.

The fact that weak interactions can involve electric charge would seem to hint at some connection between the weak and electromagnetic forces. This suggests formulating the electroweak unification theory. Just as electromagnetism unified electricity and magnetism, the electroweak model unifies electromagnetism with the weak force. Thus, the *W*'s, *Z*, and photon are all related as electroweak gauge bosons.[3] This unity only becomes significant at very high temperatures, such as during the early moments of the Big Bang, when particles and gauge bosons were massless. As the Universe cooled down, at a certain critical temperature of about 235 GeV/k_B (T is given by $E = k_B T$ where $k_B = 0.86 \times 10^{-13}$ GeV / K is Boltzmann's constant and the temperature is 273.26×10^{13} K) the *W's* and *Z* acquired mass via the Higgs mechanism and the electroweak force froze out, broke down into two pieces as it were, namely weak and electromagnetic.

7.3.2 The Higgs boson

There is a popular understanding which attributes the Higgs boson the creation of mass in particles. The Higgs boson does not technically give other particles mass. More precisely, this boson is a quantised manifestation of a

TABLE 7.3 Force fields, their interactions, and boson carriers including spin

FIELD	*SPIN*	*QUANTUM*	*INTERACTS WITH*
Gravity	2	Graviton	Everything
EM	1	Photon	Charged fields
Residual strong	0	Pion (π^+, π^-, π^0)	Baryons
Weak	1	Weakon (W^+, W^-, Z)	Leptons and baryons
Higgs	0	Higgs boson	Everything

Note that the graviton is predicted by string theory and has not been observed nor is it part of the Standard Model. The name weakon exists but is very rarely mentioned. The Higgs boson has been observed with a mass of 125.35 GeV c^{-2}.

field, the Higgs field, that generates mass through its interaction with other particles. But why could not mass just be assumed as a given?

The answer goes back to previous work in quantum field theory. Quantum fields are similar to more familiar fields, like electric and magnetic fields. But quantum fields contain excited states that we observe as particles. These fields can be divided into *matter fields* (whose particles are electrons, quarks, etc.) and *force fields* (whose particles are photons, gluons, etc.).

However, the theory had trouble modelling nuclear interactions. The short range of the weak nuclear force implied that its corresponding particles had mass, in contrast with the massless photon, the particle associated with electromagnetic fields. Simply sticking a mass onto a force-carrying particle had disastrous effects, causing certain predictions to diverge to infinity.

The solution formulated by Peter Higgs, Francois Englert, and Robert Brout proposes that the Universe is filled with a field that interacts with the weak force particles, namely W^+, W^-, and Z^0, to give them mass. It does so because the field is assumed not to be zero in empty space. This nonzero ground state violates a symmetry[4] that is considered fundamental to quantum field theory. Earlier work had shown that this kind of symmetry-breaking[5] led to a massless, spinless particle that was ruled out by experiments. Englert, Brout, and Higgs showed how one could make this unwanted particle disappear by coupling the space-filling field to the weak force field. When they worked out all of the interactions, they found that the force particles effectively had a mass, and the unwanted, massless, spinless particle was essentially absorbed by the weak particles. These particles gained a third spin state as a result, and the only remaining spinless particle was the massive Higgs boson. The Higgs boson is the visible manifestation of the Higgs field, rather like a wave at the surface of the sea.[5]

Subsequent work showed that the Brout-Englert-Higgs mechanism (or "Higgs mechanism", for short) could give mass not only to weak particles but also to fermions via what is known as the Yukawa coupling discussed in Appendix C. The more strongly a particle interacts with the Higgs field, the more massive it is. It is important to note, however, that most of the mass in composite particles, like protons, nuclei, and atoms, does not come from the Higgs mechanism, but from the binding energy that holds these particles together and kinetic energies as shown in Figure 7.2.

NOTES

1 This is a quantum effect whereby, for example, ν_μ can turn into ν_τ.
2 Which in some sense it can be likened to the valence electrons in an atom.
3 In Appendix C we show how this comes about.
4 We discuss this in Chapter 8 and Appendix B.
5 https://home.cern/science/physics/origins-brout-englert-higgs-mechanism.

PART III

Elementary particles and the first 375,000 years of the universe

Thermal history of the Universe and beyond

8

8.1 DEFINITION OF GRAND UNIFIED THEORIES

The huge amount of energy released at the start of the Big Bang resulted in the emission of ultra-relativistic particles and photons. They can be labelled on the same footing as radiation. Although the temperature reached is overloaded with uncertainty, it can be estimated. It is here where Grand Unified Theories (GUTs) of electroweak and strong interactions come into play.

The idea behind GUT is to describe the fundamental forces of the Standard Model in terms of a single unified interaction. With all players being massless, all the symmetries of the Standard Model will be present. A further symmetry should be in the strength of the interactions, otherwise the particles could not be exchanged: for some energy scale the running coupling constants must converge. To see this, let us note the following. Charged particles have virtual quantum allowed clouds of photons and electron-positron pairs around them. Coloured particles have virtual gluons and quark-antiquark pairs. As a result, the total coupling at long distance is different from the coupling at short distance where the cloud is penetrated. Electromagnetic coupling through the fine structure constant increases from 1/137 to 1/40 at around a temperature of the order of 10^{15} GeV. However, the strong force coupling constant drops from 1 to 1/40 for the same scale. The weak force coupling constant increases too as the temperature becomes

DOI: 10.1201/9781003099581-12

10^{15} GeV. The end result is that couplings come together for this temperature which is called the *unification scale*. This is illustrated in Figure 8.1.

Unification means that at a GUT scale, where masses can be ignored, all fundamental particles appear in the same multiplet, a technical term which at this stage we can only understand intuitively. This allows their charges to be the same or given fractions of each other, and accounts for the proton and electron charge being equal. In fact, in the GUT, there are vector bosons, like the weak bosons, that take fundamental particles into other fundamental particles. It is interesting to point out that the first unification was between electricity and magnetism expressed through Maxwell's equations. More recently, as we mentioned in Chapter 7, electromagnetism and weak interactions were unified to form electroweak theory.

Having reached this point, we are in a position to discuss the thermal history of the Universe.

8.2 THERMAL HISTORY OF THE UNIVERSE

In this section we concentrate on the thermal development of the early Universe from the Big Bang until the release of the Cosmic Microwave

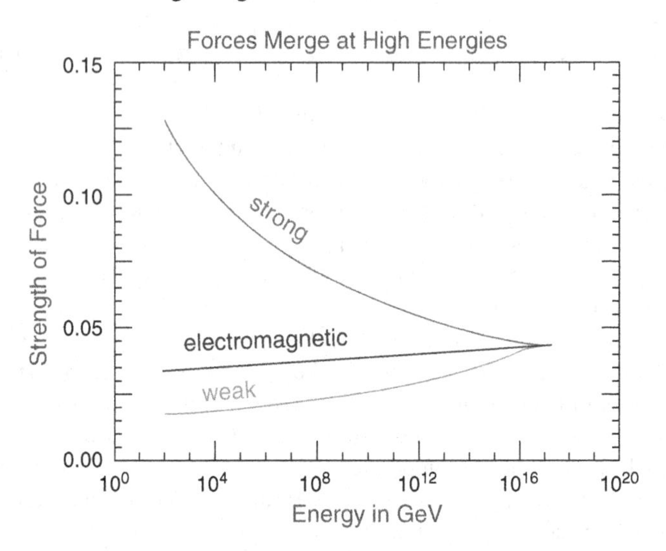

FIGURE 8.1 Strength of coupling constants, a measure of the intensity of the fundamental forces as a function of energy (temperature). In this supersymmetric description there is an exact convergence at a 10^{16} GeV energy scale. In the Standard Model the convergence is close but not exact at the same energy scale.

Background (CMB). This is a rather complex history that involves thermal equilibrium and non-equilibrium. It involves symmetry breakings by way of phase transitions and *nucleosynthesis*, namely the formation of the lightest nuclei and the corresponding atoms. These are all but a few phenomena spanned over a period of some 375,000 years of Universe history.

The symmetry introduced by the zero mass of the particles at the time of the Big Bang and the ultra-relativistic energies involved make plausible the assumption that this was a brief GUT period. The required convergence with temperature of the three coupling constants around 10^{15} GeV indicates that probably this was the GUT temperature. The simultaneous creation of antiparticles led to a thermodynamic equilibrium. As an example, we illustrate with the Feynman diagram of Figure 8.2 an electron-positron annihilation, creating photons. Of course, this process in principle applies to all particles and their antiparticles.

The key to understanding this thermal equilibrium is a comparison between the rate of interactions Γ and the rate of expansion H along the lines of our reheating discussion. This leads to the definition of the interaction characteristic time $t_c \equiv \Gamma^{-1}$ and the expansion time $t_H \equiv H^{-1}$. Thermal equilibrium for a process as in Figure 8.1 requires that $t_c \ll t_H$. According to a simple analysis it can be shown that

$$\frac{t_c}{t_H} \sim \frac{T}{10^{16}\ \text{GeV}}$$

Below a temperature of 10^{16} GeV but above 100 GeV the condition for thermal equilibrium is satisfied.

If this balance of particle-antiparticle creation and annihilation had been sustained, the Universe will have consisted of radiation only as massive particles would be suppressed. It is therefore, essential to understand that deviations from equilibrium led to the subsequent development of the Universe. This deviation leads to the idea of freeze-out of massive particles as we explain below through the electroweak freeze-out[1] when the gauge bosons acquired mass.

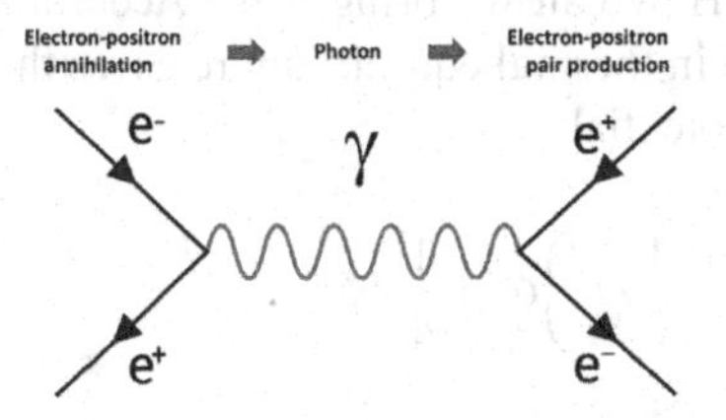

FIGURE 8.2 Electron-positron annihilation creates a photon which in turn decays into a new pair. This Feynman diagram symbolises the true reactions[2]: $e^- + e^+ \rightarrow \gamma \rightarrow e^- + e^+$.

8.3 TIMELINE

We know, of course, that matter has predominated over antimatter. This is a puzzle. The solution to it is embodied in a process called *baryogenesis*. To fix ideas, we must introduce some concepts first.

8.3.1 Baryogenesis

There is a quantum number called the *baryon number* symbolised by B. For quarks $B = 1/3$ and for antiquarks it is $-1/3$. Therefore, for baryons like the protons and neutrons $B = 1$, but for mesons $B = 0$. In any interaction the number of baryons is conserved. In fact, the number of baryons must be the same as the number of anti-baryons. If this had always been the case, then baryonic matter, as we know it, would have never come into existence. At the GUT high-energy scale, we can conceive of a process to create baryonic imbalance called *baryogenesis*. If in GUT quarks can turn into leptons within the same multiplet, and conversely, then it is possible that through some non-equilibrium mechanism, the number of baryons was not conserved in the GUT short-lived epoch: a candidate for this mechanism is that decays of the super-heavy GUT Majorana neutrinos[3] generated the imbalances between baryon and anti-baryon numbers. This is a qualitative, rather simplified argument, which nonetheless should give a hint as to how baryogenesis was triggered at a very early stage of Big Bang cosmology.

8.3.2 Electroweak symmetry breaking

Symmetries that are broken in the present Universe were unbroken in the early Universe. This is known as *symmetry restoration*, i.e., in what we are considering here, the symmetry restoration occurs via a sufficient increase in the temperature. How can we bring it in? According to Quantum Field Theory any scalar ϕ in thermal equilibrium receives the following contribution to its effective potential:

$$V(\phi,T) = \left(gT^2 - \frac{1}{2}m^2\right)\phi^2 + \frac{1}{4}\lambda\phi^4 \quad .$$

where g is the coupling constant between ϕ and the thermal reservoir. In Figure 8.3 we plot $V(\phi,T)$ for different temperatures. We notice that, above a critical temperature T_c, the minimum occurs at $\phi = 0$ which gives full symmetry. As $T < T_c$ the symmetry is spontaneously broken with the appearance of the two minima.

The scalar field of Figure 8.3 is the Higgs field, which drives the electroweak phase transition at a temperature reached (200 GeV) one nanosecond after the Big Bang. Just prior to this transition, particle interaction timescales are far shorter than the timescale of the evolving Universe, and the radiation which fills the Universe is, in effect, in perfect thermal equilibrium. Figure 8.3 shows the phase transition from full symmetry to spontaneous symmetry breaking as the Universe cools down through its expansion. The first phase transition is the *electroweak symmetry breaking* which occurs at $T_c \approx 200$ GeV when the gauge bosons $W^{\mp}$ and Z acquire mass. It is said that the electroweak force freezes out of the Grand Unification leaving "behind" a quark-gluon plasma.

8.3.3 Quantum Chromodynamics (QCD) and the quark-gluon phase transition

As Figure 8.1 shows, at high temperatures, or equivalently at high energy scales, the strong force coupling constant becomes weaker. As a result of this weakened interaction between quarks, hadronic matter is in a state of a quark-gluon plasma.

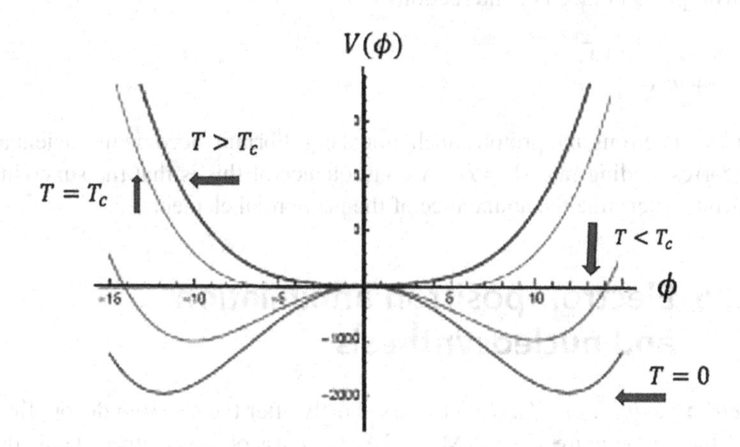

FIGURE 8.3 Effective potential $V(\phi, T)$ for four different temperatures relative to the critical temperature $T_c = m / g\sqrt{2}$ at which a phase transition occurs. For $T < T_c$ the symmetry is broken and for $T > T_c$ the symmetry is restored.

For an energy scale $\Lambda_{\text{QCD}} \sim 200\,\text{MeV}$, $\alpha_s\left(T^2\right)$ acquires significance to cause a QCD *phase transition*: quarks and gluons form hadrons, namely baryons and mesons. Prominent baryons are the proton and the neutron, whereas pions are the prominent mesons.

8.3.4 Neutrino decoupling

The next milestone is *neutrino decoupling*. Neutrinos are coupled to the thermal bath via the weak interaction, e.g.,

$$\nu_e + \overline{\nu_e} \leftrightarrow e^+ + e^-$$

It can be shown that the ratio of the interaction rate Γ and the Hubble rate H is

$$\frac{\Gamma}{H} = \left(\frac{T}{1\,\text{MeV}}\right)^3$$

Thus, the rate of interaction drops fast with temperature with the decoupling occurring at $T \sim 1\,\text{MeV}$. When this happens the neutrinos move freely along geodesics (which we defined earlier in the book). The decoupling of the primordial neutrinos from other particles is correlated with the freeze-out of the neutron-proton ratio. The interactions,

$$n \leftrightarrow p + e + \overline{\nu}_e$$
$$n + \nu_e \leftrightarrow p + e$$

that kept neutrons and protons in chemical equilibrium, become inefficient as the corresponding rates $\Gamma = H$. A consequence of this is that the surviving neutrons determine the abundance of the primordial elements.

8.3.5 Electron-positron annihilation and nucleosynthesis

Electron-positron annihilation occurs shortly after the neutrino decoupling. Here the temperature $T \sim 0.5\,\text{MeV}$, i.e., the mass of the electron. Thus, the equilibrium in electron-positron annihilation expressed as

$$e^+ + e^- \leftrightarrow \gamma + \gamma$$

cannot back-convert from photons when the temperature drops below the electron mass. Only a small fraction of electrons survive, about 1 per 10^9 photons, to maintain electrical neutrality with the protons. Then the production of photons raises the temperature of the photons that remain in thermal equilibrium.

When the temperature reaches $T \sim 0.05$ MeV nuclear reactions become efficient and free protons and neutrons interact to form light nuclei like deuterium and helium. This is the stage of *nucleosynthesis.*

8.3.6 Recombination and the CMB

In the early Universe, at temperatures $T \geq$ eV, the atoms in the Universe are ionised, and photons are tightly coupled to the baryons through Thomson scattering from the electrons. At a temperature $T \sim$ eV, electrons and protons combine to form hydrogen atoms, free electrons disappear, and without any electrons to scatter from, photons decouple and free stream through the Universe. These are the CMB photons we see today.

A major event in the thermal history of the Universe was, when electrons coupled to protons to form atoms. When the temperature was $T \geq 1$ eV, the Universe consisted of plasma of electrons and nuclei. Photons were tightly coupled to the plasma via Thomson scattering. Within the plasma, the dominant interaction was Coulomb scattering. When the temperature dropped due to expansion, electrons and nuclei combined to form neutral atoms. Photon scattering dropped to such an extent that the mean free path of photons became larger than the Hubble radius. *Decoupling* of photons from matter was the result and the Universe became transparent as illustrated by Figure 8.4. These photons roamed through space maintaining a blackbody radiation thermal equilibrium which we know today as CMB.

BOX 8.1: SURFACE OF LAST SCATTER

CMB radiation's "surface of last scatter" is analogous to the light coming through the clouds to our eye on a cloudy day. When the Wilkinson Microwave Anisotropy Probe (WMAP) observes the microwave background sky it looks back to when there were free electrons that could readily scatter cosmic background radiation. This cosmic background "cloud surface" is called the "surface of last scatter". If there were any "features" imprinted in this surface of last scatter (i.e. regions that were brighter or dimmer than average) they will remain imprinted to this day because emitted light travels across the Universe largely unimpeded.

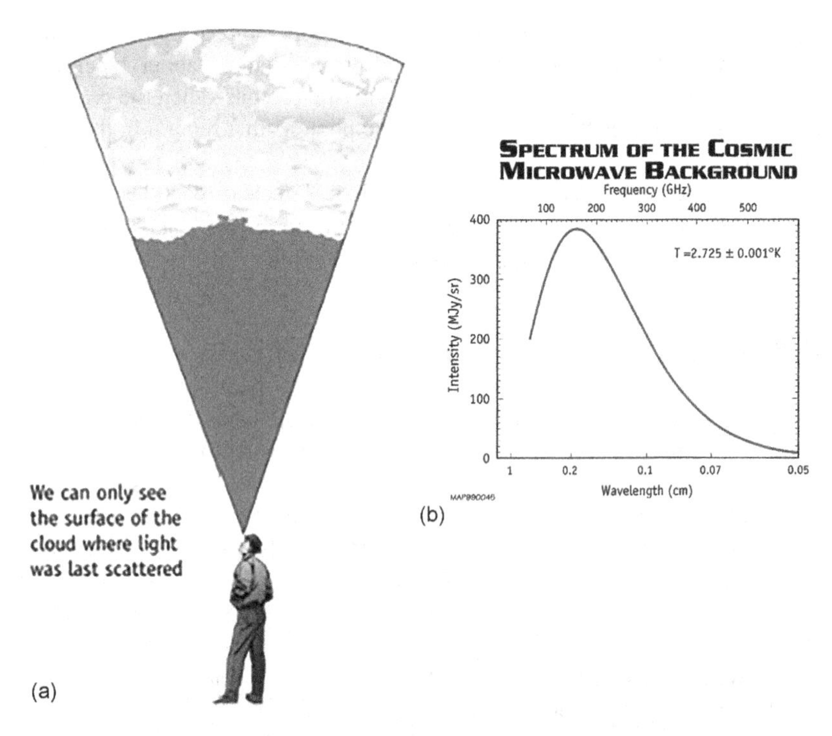

FIGURE 8.4 Before recombination the Universe was opaque due to Thomson scattering. When atoms formed, the Universe became transparent (a) Illustration of the surface of last scatter by comparing it with the surface of a cloud where the light was last scattered. (b) Theoretical perfect fit for the spectrum of CMB. This was done by applying Wien's law of blackbody radiation (Courtesy: NASA).

Finally, in Figure 8.5, we show a graphical synopsis.

NOTES

1. Freeze-out means dropping out of GUT and having an "independent" existence.
2. If the e^-e^+ are at rest or the total momentum is 0, two photons are produced with opposite momenta.
3. They are massive Majorana fermions, i.e., they are their own antiparticles.

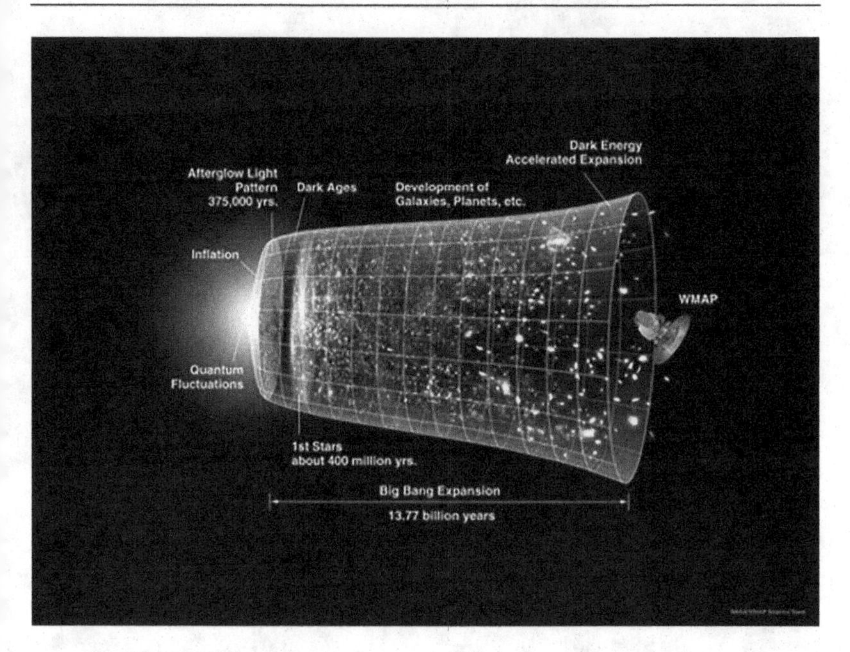

FIGURE 8.5 A representation of the evolution of the Universe over 13.77 billion years. The far left depicts the earliest moment we can now probe, when a period of "inflation" produced a burst of exponential growth in the Universe. (Size is depicted by the vertical extent of the grid in this graphic.) For the next several billion years, the expansion of the Universe gradually slowed down as the matter in the Universe pulled on itself via gravity. More recently, the expansion has begun to speed up again as the repulsive effects of dark energy have come to dominate the expansion of the Universe. The afterglow light seen by WMAP was emitted about 375,000 years after inflation and has traversed the Universe largely unimpeded since then. The conditions of earlier times are imprinted on this light; it also forms a backlight for later developments of the Universe (Courtesy: NASA).

Appendix A

Elements of group theory

We present a rudimentary introduction to aid understanding of gauge symmetries. Group theory plays a fundamental role in particle physics.[1] The reason it does so is because group theory is related to symmetry. For example, an important group in physics is the rotation group, which is related to the fact that the laws of physics do not change if we rotate the frame of reference. In general, what we are after is to make sure that equations keep the same mathematical form under various transformations. Group theory is related to this type of symmetry.

A.1 WHEN DOES A SET OF ELEMENTS FORM A GROUP?

What is a group? A group G is a set of elements $\{a, b, c, \ldots\}$ which includes a multiplication or composition rule such that if $a, b \in G$, then the product is also a member of the group, that is,

$$ab \in G \tag{A.1}$$

A group must satisfy four axioms, namely,

1. Associativity: the multiplication rule is associative $(ab)c = a(bc)$.
2. Identity element: the group has an identity element e that satisfies $ae = ea = a$. The identity element for a group is unique.
3. Inverse element: for every element $a \in G$, there exists an inverse which we denote by a^{-1} such that $a^{-1}a = aa^{-1} = e$.
4. Order: the order of the group is the number of elements that belong to G.

Let us put into practice the above axioms to show how they work to define whether a set forms a group.

Example 1: Set of integers under addition

The set of all integers form a group under addition. We can take the composition rule to be addition. Let $z_1, z_2 \in Z$. Clearly the sum

$$z_1 + z_2 \in Z$$

is another integer, so it belongs to the group. For the identity, we can take $e = 0$ since

$$z + 0 = 0 + z = z$$

for any $z \in Z$. Addition is commutative, that is,

$$z_1 + z_2 = z_2 + z_1$$

Because of this commutation property, the group is said to be *Abelian*. The inverse of z is just $-z$ since

$$z + (-z) = e = 0$$

which satisfies $aa^{-1} = a^{-1}\ a = e$.

A.2 REPRESENTATION OF A GROUP

In particle physics we are often interested in what is called a *representation* of the group. Let us denote that representation by Γ. Representation is a mapping that takes group elements $g \in G$ into linear operators that preserve the composition rule of the group in the sense that,

- $\Gamma(a)\Gamma(b) = \Gamma(ab)$
- The representation also preserves the identity, that is, $\Gamma(e) = I$

A.3 LIE GROUPS

An ordinary function of position is specified by an input x to get $y = f(x)$. In a similar way, a group can also be a function of one or more inputs that we call *parameters*. Let the group G be such that individual elements $g \in G$ are specified by a finite set of parameters, say n of them. If we denote the set of parameters by

$$\{p_1, p_2, \ldots, p_n\}$$

then the group element is written as

$$g = G(p_1, p_2, \ldots, p_n)$$

The identity is the group element where the parameters are all set to 0

$$e = G(0, 0, \ldots, 0)$$

While there are discrete groups with a finite number of elements, most of the groups we will be concerned with have an infinite number of elements. However, they have a finite set of continuously varying parameters.

In the expression $g = G(p_1, p_2, \ldots, p_n)$, for our purposes the parameters will be angles α_k, since several important groups in physics are related to rotations. The angles vary continuously over the finite range $0 \le \alpha_k < 2\pi$. In addition, the group is parameterised by a finite number of parameters, the angles of rotation.

So, if a group G

- Depends on a finite set of continuous parameters α_k
- the derivatives of the group elements with respect to all parameters exist,

we call the group a *Lie Group*. To simplify the discussion, we begin with a group with a single parameter α. We obtain the identity element by setting $\alpha = 0$. Thus,

$$g(\alpha)\big|_{\alpha=0} = e \tag{A.2}$$

By taking derivative with respect to the parameter and evaluating the derivative at $\alpha = 0$, we obtain the *generator* of the group. Let us denote an abstract generator by Q. Then,

$$Q = \left.\frac{\partial g}{\partial \alpha}\right|_{\alpha=0} \tag{A.3}$$

More generally, if there are n parameters in the group, then there will be n generators such that a generator Q_k is given by

$$Q_k = \left.\frac{\partial g}{\partial \alpha_k}\right|_{\alpha_k=0} \tag{A.4}$$

Rotations have a special property in that they are length-preserving: rotate a vector, and it maintains the same length. Rotation by $-\alpha$ cancels the rotation by α, hence rotations have an *orthogonal* or *unitary* representation. In the case of quantum theory, we seek a unitary representation of the group and choose the generators Q_i to be Hermitian. In this case we replace, $H_k \equiv -iQ_k$

$$H_k = -i\left.\frac{\partial g}{\partial \alpha_k}\right|_{\alpha_k=0} \tag{A.5}$$

For some finite α, the generators allow us to define a representation of the group. Consider a small number > 0 and use Taylor expansion to form a representation of the group which we denote by D,

$$D(\ \alpha) = 1 + \ \alpha H$$

If $\alpha = 0$, then clearly the representation gives the identity. It is also worth recalling the series expansion of the exponential function

$$e^x = 1 + x + \frac{x^2}{2!} + \frac{x^3}{3!} + \cdots$$

So we can define the representation of the group in terms of the exponential using the mathematical identity

$$D(\theta) = \lim_{n\to\infty}\left(1 + i\frac{\alpha H}{n}\right)^n = e^{i\alpha H} \tag{A.6}$$

Notice that if H is Hermitian, then $H = H^\dagger$ and the representation of the group is unitary, since

$$D^\dagger(\alpha) = \left(e^{i\alpha H}\right)^\dagger = e^{-i\alpha H}$$

$$\Rightarrow D^\dagger(\alpha) D(\alpha) = e^{-i\alpha H} e^{i\alpha H} = 1$$

One reason that the generators of the group are important is that they form a vector space. This means we can add two generators of the group together to obtain a third generator, and we can multiply generators by a scalar, and still have a generator of the group. A complete vector space can be used as a basis for representing other vector spaces. For example, the Pauli matrices, from quantum mechanics, can be used to describe any 2×2 matrix.

The character of a group is defined in terms of the generators in the following sense. The generators satisfy a commutation relation which we write as

$$[H_m, H_k] \equiv H_m\ H_k - H_k\ H_m = i f_{mkq}\ H_q \tag{A.7}$$

where typically

$$f_{mkq} = \begin{cases} 0 & \text{if any two subscripts are equal} \\ +1 & \text{if the subscripts form an even permutation} \\ -1 & \text{if the subscripts form an odd perrmutation} \end{cases}$$

Equation (A.7) is called the *Lie algebra* of the group. The quantities f_{mkq} are the *structure constants* of the group. Looking at the commutation relation for the Lie algebra, Eq. (A.7), it is a reminder of the Pauli matrices which we will come back to later on.

A.4 THE ROTATION GROUP

The rotation group is the set of all rotations about the origin. A key feature is that rotations preserve the lengths of vectors. This mathematical property is expressed by saying the matrices are unitary. It is easy to see that the set of rotations forms a group.

Rotations satisfy the group composition rule. If R_1 and R_2 are two rotations, then we can say that

$$R_3(\theta_3) = R_2(\theta_2)R_1(\theta_1)$$

Two successive rotations θ_1 and θ_2 are equivalent to a single rotation θ_3. Hence R_3 is also a rotation that belongs to the group.

In general rotations do not commute. That is,

$$R_1R_2 \neq R_2R_1$$

However, rotations are associative

$$R_1(R_2R_3) = (R_1R_2)R_3$$

The rotation group has an identity element which is simply doing no rotation at all. The inverse of a rotation is simply the rotation carried out in the opposite direction,

$$R(-\theta)R(\theta) = I$$

A.5 HOW TO REPRESENT ROTATIONS?

Suppose we have a rotation in the plane. A point in the plane will have coordinates which we can call x_1 and x_2. Now let us perform a rotation by an angle θ with respect to the x-axis. Although the point stays the same, that is, in the same place in the plane, the rotated frame will cause this point to have different coordinates. The relationship between the old coordinates and the new coordinates

$$\begin{pmatrix} x' \\ x_2' \end{pmatrix} = \begin{pmatrix} \cos\theta & \sin\theta \\ -\sin\theta & \cos\theta \end{pmatrix} \begin{pmatrix} x_1 \\ x_2 \end{pmatrix} = \begin{pmatrix} x_1\cos\theta + x_2\sin\theta \\ -x_1\sin\theta + x_2\cos\theta \end{pmatrix}$$

The rotation is represented by the matrix,

$$R(\theta) \equiv \begin{pmatrix} \cos\theta & \sin\theta \\ -\sin\theta & \cos\theta \end{pmatrix}$$

The transpose of the rotation matrix is

$$R^T(\theta) \equiv \begin{pmatrix} \cos\theta & -\sin\theta \\ \sin\theta & \cos\theta \end{pmatrix}$$

Then

$$R(\theta)R^T(\theta) = \begin{pmatrix} \cos\theta & \sin\theta \\ -\sin\theta & \cos\theta \end{pmatrix}\begin{pmatrix} \cos\theta & -\sin\theta \\ \sin\theta & \cos\theta \end{pmatrix} = \begin{pmatrix} 1 & 0 \\ 0 & 1 \end{pmatrix}$$

And the identity is

$$I = \begin{pmatrix} 1 & 0 \\ 0 & 1 \end{pmatrix}$$

This tells us that the transpose of the matrix is the inverse group element, since multiplying two matrices together gives the identity.

A.6 SPECIAL ORTHOGONAL GROUP SO(*N*)

The group SO(*N*) are special orthogonal $N \times N$ matrices. The term *special* is a reference to the fact that these matrices have determinant +1. A larger group, one that contains SO(*N*) as a subgroup, is the orthogonal group O(*N*) composed of orthogonal $N \times N$ matrices that can have arbitrary determinant. Generally speaking, rotations can be represented by orthogonal matrices, which themselves form a group. So the group SO(3) is the representation of rotations in three dimensions, and the group consists of 3×3 orthogonal matrices with determinant +1.

Consider the familiar case of SO(3). This group has three parameters, the three angles defining their rotations about the three axes. Let these angles be denoted by ξ, ϕ and θ. Then the rotations about *x*, *y*, and *z* are, respectively,

$$R_x(\xi) = \begin{pmatrix} 1 & 0 & 0 \\ 0 & \cos\xi & \sin\xi \\ 0 & -\sin\xi & \cos\xi \end{pmatrix}$$

$$R_y(\phi) = \begin{pmatrix} \cos\phi & 0 & \sin\phi \\ 0 & 1 & 0 \\ -\sin\phi & 0 & \cos\phi \end{pmatrix}$$

$$R_z(\theta) = \begin{pmatrix} \cos\phi & \sin\theta & 0 \\ -\sin\theta & \cos\theta & 0 \\ 0 & 0 & 1 \end{pmatrix}$$

These matrices are the representation of rotations in three dimensions. Rotations in three or more dimensions do not commute, however. It is an easy although tedious exercise to show that the rotation matrices written down here do not commute either. The task now is to find the generators J_x, J_y, J_z for each group parameter. We do this using Eq. (A.5):

$$J_x = -i\frac{dR_x}{d\xi}\bigg|_{\xi=0} = \begin{pmatrix} 0 & 0 & 0 \\ 0 & 0 & -i \\ 0 & i & 0 \end{pmatrix}$$

$$J_y = -i\frac{dR_y}{d\phi}\bigg|_{\phi=0} = \begin{pmatrix} 0 & 0 & -i \\ 0 & 0 & 0 \\ i & 0 & 0 \end{pmatrix}$$

$$J_z = -i\frac{dR_y}{d\theta}\bigg|_{\theta=0} = \begin{pmatrix} 0 & -i & 0 \\ i & 0 & 0 \\ 0 & 0 & 0 \end{pmatrix}$$

These matrices are the familiar angular momentum matrices. This agrees with the well-known result that the angular momentum operators are the generators of rotations.

For example, an infinitesimal rotation about the z-axis by an angle $\varepsilon\theta$, where ε is a small positive parameter, is given by

$$R_z(\varepsilon\theta) = 1 + iJ_z\varepsilon\theta$$

where we need the imaginary unit to make the rotation real. To obtain a rotation by a finite angle θ, we would have to apply the above rotation an infinite number of times. At the same time, we must guarantee that the quantity remains infinitesimal in terms of the angle as we apply successively the rotation matrix an infinite number of times. We can sum up this as follows:

$$R_z(\theta) = (1 + iJ_z\varepsilon\theta)(1 + iJ_z\varepsilon\theta)\ldots(1 + iJ_z\varepsilon\theta)\ldots\ n \text{ times with } n \gg 1$$

More accurately, we also must keep ε smaller and smaller as $n \to \infty$, thus

$$R_z(\theta) = \lim_{n\to\infty}\left(1 + \frac{iJ_z\varepsilon\theta}{n}\right)^n = e^{iJ_z\theta}$$

Since the exponent is a matrix, these results in principle are not defined. However, it should be interpreted in terms of the series expansion, because then we will get powers of the matrix which means that the matrix must be applied successively as the number indicated by the power. Thus,

$$e^{iJ_z\theta} = 1 + iJ_z\theta + \frac{(iJ_z\theta)^2}{2!} + \frac{(iJ_z\theta)^3}{3!} + \frac{(iJ_z\theta)^4}{4!} + \cdots$$

Therefore,

$$e^{iJ_z\theta} = 1 + iJ_z\theta - \frac{(J_z\theta)^2}{2!} - i\frac{(J_z\theta)^3}{3!} + \frac{(J_z\theta)^4}{4!} +$$

Note the following,

$$J_z^2 = \begin{pmatrix} 0 & -i & 0 \\ i & 0 & 0 \\ 0 & 0 & 0 \end{pmatrix}\begin{pmatrix} 0 & -i & 0 \\ i & 0 & 0 \\ 0 & 0 & 0 \end{pmatrix} = \begin{pmatrix} 1 & 0 & 0 \\ 0 & 1 & 0 \\ 0 & 0 & 0 \end{pmatrix}$$

It then follows that for even powers,

$$J_z^2 = J_z^4 = J_z^6 = \cdots = \begin{pmatrix} 1 & 0 & 0 \\ 0 & 1 & 0 \\ 0 & 0 & 0 \end{pmatrix}$$

For odd powers,

$$J_z^3 = J_z J_z^2 = \begin{pmatrix} 0 & -i & 0 \\ i & 0 & 0 \\ 0 & 0 & 0 \end{pmatrix}\begin{pmatrix} 1 & 0 & 0 \\ 0 & 1 & 0 \\ 0 & 0 & 0 \end{pmatrix} = \begin{pmatrix} 0 & -i & 0 \\ i & 0 & 0 \\ 0 & 0 & 0 \end{pmatrix} = J_z$$

Thus,

$$J_z = J_z^3 = J_z^5 = J_z^7 = \cdots = \begin{pmatrix} 0 & -i & 0 \\ i & 0 & 0 \\ 0 & 0 & 0 \end{pmatrix}$$

Now we can group the terms together as shown by the curly brackets:

$$e^{iJ_z\theta} = \left\{ i\begin{pmatrix} 0 & -i & 0 \\ i & 0 & 0 \\ 0 & 0 & 0 \end{pmatrix}\theta - i\begin{pmatrix} 0 & -i & 0 \\ i & 0 & 0 \\ 0 & 0 & 0 \end{pmatrix}\frac{(\theta)^3}{3!} + \ldots \right\}$$

$$+\left\{ \begin{pmatrix} 1 & 0 & 0 \\ 0 & 1 & 0 \\ 0 & 0 & 0 \end{pmatrix} - \begin{pmatrix} 1 & 0 & 0 \\ 0 & 1 & 0 \\ 0 & 0 & 0 \end{pmatrix}\frac{(\theta)^2}{2!} + \begin{pmatrix} 1 & 0 & 0 \\ 0 & 1 & 0 \\ 0 & 0 & 0 \end{pmatrix}\frac{(\theta)^4}{4!} + \ldots \right.$$

$$= \begin{pmatrix} 0 & \theta & 0 \\ -\theta & 0 & 0 \\ 0 & 0 & 0 \end{pmatrix} - \begin{pmatrix} 0 & \frac{(\theta)^3}{3!} & 0 \\ -\frac{(\theta)^3}{3!} & 0 & 0 \\ 0 & 0 & 0 \end{pmatrix} + \ldots \begin{pmatrix} 1 & 0 & 0 \\ 0 & 1 & 0 \\ 0 & 0 & 0 \end{pmatrix}$$

$$- \begin{pmatrix} \frac{(\theta)^2}{2!} & 0 & 0 \\ 0 & \frac{(\theta)^2}{2!} & 0 \\ 0 & 0 & 0 \end{pmatrix} + \begin{pmatrix} \frac{(\theta)^4}{4!} & 0 & 0 \\ 0 & \frac{(\theta)^4}{4!} & 0 \\ 0 & 0 & 0 \end{pmatrix} + \ldots$$

We now gather all the terms in one large matrix to see what it tells us

$$e^{iJ_z\theta} = \begin{bmatrix} \left(1 - \frac{\theta^2}{2!} + \frac{\theta^4}{4!}\right) & \left(\theta - \frac{\theta^3}{3!}\right) & 0 \\ -\left(\theta - \frac{\theta^3}{3!}\right) & \left(1 - \frac{\theta^2}{2!} + \frac{\theta^4}{4!}\right) & 0 \\ 0 & 0 & 0 \end{bmatrix} \tag{A.8}$$

Now consider the following Taylor expansions,

$$\sin\theta = \theta - \frac{\theta^3}{3!} + \ldots$$

$$\cos\theta = 1 - \frac{\theta^2}{2!} + \ldots$$

Although these are truncated expansions, it is very clear that the relevant elements that appear in the matrix of Eq. (A.8) are identical with the expansion in Eq. (A.9). Substitution in Eq. (A.8) will give us the well-known expression for rotation around the *z*-axis and therefore it gives us a simple operational representation of the rotation.

A.7 SPECIAL UNITARY GROUP SU(2)

In particle physics unitary groups play a special role. This is so since unitary operators play an important role in quantum theory. Specifically, unitary operators *preserve the inner product* $<\psi|\phi\rangle$, which means that the unitary transformation leaves the probabilities for different transition among the states unaffected. That is, quantum physics is invariant under a unitary transformation.

When the physical predictions of a theory are invariant under the action of some group, we can represent the group by a unitary operator *U*. The unitary group U(*N*) consists of all $N \times N$ unitary matrices. Special unitary groups, denoted by SU(*N*), are $N \times N$ unitary matrices with positive unit determinant. The dimension of SU(*N*), hence the number of generators, is given by $N^2 - 1$. We are particularly interested in SU(2), which is applicable in the context of electroweak theory discussed in Appendix C.

The rank gives the number of operators in the algebra that can be simultaneously diagonalised. The rank of SU(*N*) is $N - 1$. So for SU(2) we have the following:

- $2^2 - 1 = 3$ generators
- Rank $2 - 1 = 1$

The simplest unitary group is the group $U(1)$ which consists of a single complex number written in polar representation. Put it in another way, we can say that an $U(1)$ symmetry has a single parameter θ and is written as

$$U = e^{i\theta}$$

Let us take the Lagrangian of a complex field:

$$\mathcal{L} = \partial_\mu \phi^* \partial^\mu \phi - m^2 \phi^* \phi$$

which is invariant under the transformation $\phi \rightarrow e^{i\theta}\ \phi$, assuming that θ is constant. If it is not, we have to resort to gauge fields.

The unitary group $U(2)$ is the set of all 2×2 unitary matrices. Being unitary, these matrices satisfy

$$UU^{\dagger} = U^{\dagger}U = I$$

In the context of physics, we are particularly interested in a subgroup which is known as SU(2). The generators of this group are the Pauli matrices, which we reproduce here

$$\{a,\ b,\ c,...\}$$

We see now how the rank of a unitary group comes into play. The rank of SU(2) is 1, there is one diagonalised operator, σ_3, in the basis we have chosen. The generators of this group should actually be taken to be $1/2\sigma_i$ and the Lie algebra is the familiar commutation relations that are satisfied by the Pauli matrices

$$\left[\frac{\sigma_m}{2},\frac{\sigma_l}{2}\right] = i\ _{mlk}\ \frac{\sigma_k}{2}$$

where m,l,k take up the values 1, 2, or 3. U_{mlk} is the Ricci tensor

$$U_{mlk} = \begin{cases} 0 & \text{if any two indices are the same} \\ +1 & \text{if the indices are an even permutation of 123} \\ -1 & \text{if the indices are an odd permutation of 123} \end{cases}$$

A.8 SPINORS AND ROTATION

Consider the spin states of an electron. Since this is a spin-1/2 particle, there are two independent states which conventionally are termed “up” and “down”.

$$|\text{up}\rangle = \begin{pmatrix} 1 \\ 0 \end{pmatrix}\ |\text{down}\rangle = \begin{pmatrix} 0 \\ 1 \end{pmatrix}$$

The most general spin state $|\phi\rangle$ can be expanded in terms of these,

$$|\phi = c_1 \begin{pmatrix} 1 \\ 0 \end{pmatrix} + c_2 \begin{pmatrix} 0 \\ 1 \end{pmatrix}$$

How does a rotation in real space affect a spinor? Consider an infinitesimal rotation around the z-axis. Then following previous derivation,

$$R_z(\ \theta) = 1 + i\frac{\sigma_z}{2}\ \theta$$

The angular momentum operator is $S_z = \frac{\hbar \sigma_z}{2}$. Then

$$R_z(\theta) = \lim_{n\to\infty}\left(1 + \frac{i\sigma_z\ \theta}{2n}\right)^n = e^{i\sigma_z\theta/2}$$

Note that

$$\sigma_z^2 = \begin{pmatrix} 1 & 0 \\ 0 & 1 \end{pmatrix}\ \sigma_z^3 = \begin{pmatrix} 1 & 0 \\ 0 & 1 \end{pmatrix} \sigma_z = \begin{pmatrix} 1 & 0 \\ 0 & -1 \end{pmatrix}$$

Following the same steps as in the earlier derivation, the rotation matrix representation around the z-axis for a two-component spinor is

$$R_z(\theta) = \begin{pmatrix} \cos\theta & \sin\theta \\ -\sin\theta & \cos\theta \end{pmatrix}$$

In a similar fashion we can derive rotations about the other two axes.

NOTE

1 This Appendix is of interest to the mathematically inclined reader. It can be skipped without loss of continuity. We follow closely D. McMahon, *Quantum Field Theory DeMystified.*

Appendix B

The Standard Model of Particle Physics II

Spontaneous symmetry breaking

We have stressed earlier in the book that the symmetries of the Standard Model require the particles to be massless. More specifically the theory has to be *renormalisable*. This basically means that infinities arising in calculated quantities, e.g., perturbation theory in quantum electrodynamics, can be dealt with. This over-simplified definition is all we need to know in this chapter to appreciate the paramount importance of *spontaneous symmetry breaking* in the Higgs mechanism. As we show, this is partly the reason why the weak force bosons acquire mass: what is missing is gauge symmetry which we deal with in Appendix C. Fermions get their mass from the Higgs field via what is known as the Yukawa coupling.

B.1 THE FERROMAGNET

First, let us see the meaning of spontaneous symmetry breaking in a familiar example: a ferromagnet. In Figure B.1a we show the ferromagnet above a critical temperature T_c. The individual magnetic moments are oriented at random and there is no net magnetisation. As the temperature is lowered, the ferromagnet goes through a second-order phase transition[1] as it gets past T_c. The magnetic moments spontaneously acquire a particular orientation; thus, a preferred direction develops, and the symmetry is broken (Figure B.1b).

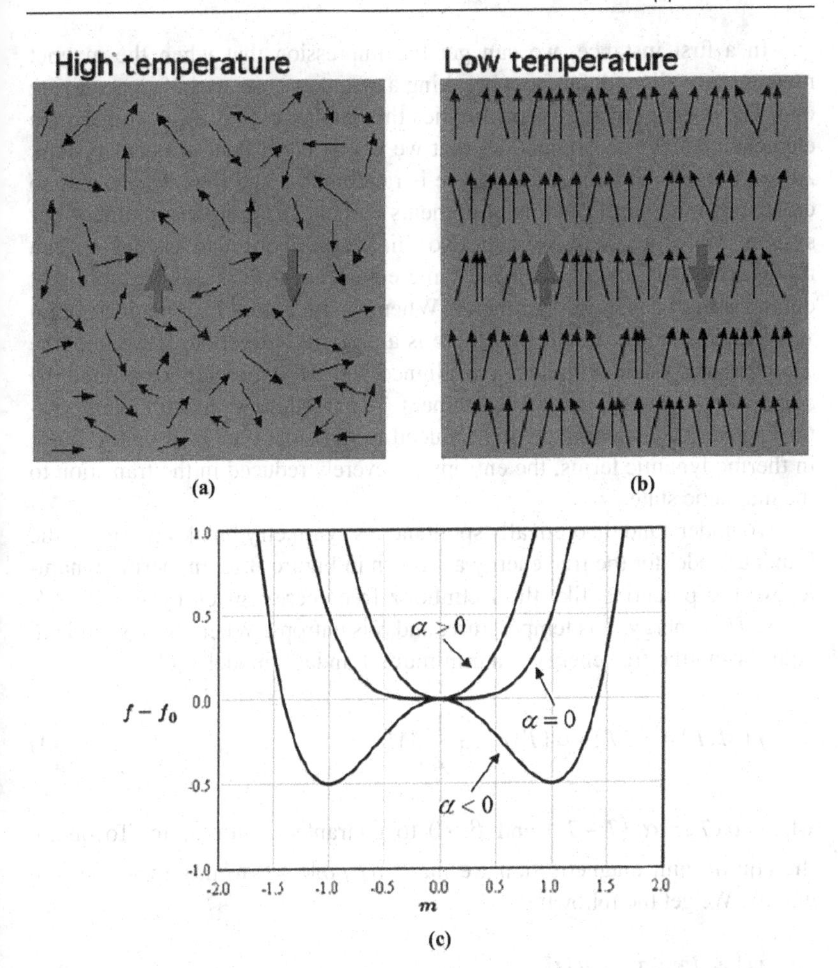

FIGURE B.1 Spontaneous symmetry breaking in a ferromagnet below a critical temperature T_c. (a) When the temperature $T > T_c$ the individual atomic magnets are oriented at random and the net equilibrium magnetisation $M = 0$. (b) When $T < T_c$ there is a spontaneous equilibrium magnetisation which breaks the symmetry as now there is a preferred direction. (c) Landau model of the free energy $f(M,T)$against M for different temperatures. Here $\alpha = \alpha_0(T - T_c)$. The curve corresponding to (a) is $\alpha > 0$ similar to a parabola and the minimum of $f(M,T)$, i.e., equilibrium, corresponds to M = 0. This is a full symmetry point as $f(M,T) = f(-M,T)$. When $\alpha < 0$ the system goes through a second-order phase transition and can settle in either of the minima where the symmetry is broken (Courtesy: Ivo van Lupen).

In a first instance, we can get the impression that when the magnet moments are aligned, far from breaking a symmetry, we are in fact acquiring one. The concept of symmetry implies the number of ways we can have the elementary magnets oriented so that we get an equivalent physical system. Above the critical temperature there is randomness and thus we are able to create any number of random alignments without changing the nature of our system. More precisely, we can also think of a coordinate system we can rotate arbitrarily and observe the same environment. It is this freedom that defines a high degree of symmetry. When we are below T_c, a magnetisation vector $\boldsymbol{M}$ develops. As a result, there is a preferred direction. If we imagine a coordinate system with its z-axis aligned with $\boldsymbol{M}$, only rotations around the z-axis will lead to the same environment. The freedom we had for a temperature above the critical is greatly reduced to only one type of rotation. Put it in thermodynamic terms, the entropy is severely reduced in the transition to the magnetic state.

To understand theoretically spontaneous symmetry breaking, we use the Landau model for the free energy as shown in Figure B.1c. In thermodynamics we use potentials like the Helmholtz free energy, given by $F = U - TS$, where U is energy, T is temperature, and S is entropy. When the system is in equilibrium the free energy is a minimum. Landau's model reads,

$$f(M,T) = f_0(T) + \alpha(T)M^2 + \frac{1}{2}\beta M^4 \tag{B.1}$$

where $\alpha(T) = \alpha_0(T - T_c)$ and $\beta > 0$ to guarantee a minimum. To obtain the equilibrium magnetisation we make $\partial f / \partial M = 0$ to minimise the free energy. We get the following:

$$M\left[\alpha_0(T - T_c) + \beta M^2\right] = 0 \tag{B.2}$$

with minima given by

$$M = M_0 = 0 \quad (T > T_c)$$

$$M_{\mp} = \mp\sqrt{\frac{\alpha_0(T_c - T)}{\beta}} \qquad (T < T_c)$$

As illustrated in Figure B.1c, when $T > T_c$ we obtain a parabola-like curve where the equilibrium point M_0 exhibits reflection symmetry $f(M) = f(-M)$. When $T < T_c$ we get a double well where either of the minima $M_{\mp}$ breaks the reflection symmetry. This is a condensed-matter analogue to understand the Higgs mechanism.

B.2 LAGRANGE FIELD THEORY

We begin by discussing the Klein-Gordon Equation (KGE), the first relativistic answer to the non-relativistic Schrödinger equation. The KGE looks the same under a transformation from a fixed frame of reference to a moving inertial one. However, it does not have the wavefunction interpretation because it can be shown that it could have negative "probability" density. Rather, the KGE is the wave equation of a field φ whose quanta are scalar particles, namely spin-0 particles. To fix ideas, just think of the electromagnetic field whose quanta are the photons.

In particle physics the Lagrangian plays a fundamental role. From classical mechanics it is defined as

$$L = K - V$$

where K is kinetic energy and V is potential energy. When we deal with a continuous scalar field φ, we need a Lagrangian density, that is, a Lagrangian per unit volume. As an example,

$$\mathcal{L} = \mathcal{K} - \mathcal{V} \tag{B.3}$$

where $\mathcal{K}$ is the kinetic energy density given by,

$$\mathcal{K} \equiv \frac{1}{2}\left(\frac{1}{c}\frac{\partial \varphi}{\partial t}\right)^2 - \frac{1}{2}(\nabla\varphi)^2 = \frac{1}{2}\left(\frac{1}{c}\frac{\partial \varphi}{\partial t}\right)^2 - \frac{1}{2}\left[\left(\frac{\partial \varphi}{\partial x}\right)^2 + \left(\frac{\partial \varphi}{\partial y}\right)^2 + \left(\frac{\partial \varphi}{\partial z}\right)^2\right] \tag{B.4}$$

and $\mathcal{V}$ is the potential energy density. Thus, the kinetic and potential energies are, respectively,

$$K = \int \mathcal{K}\ dxdydz$$

$$V = \int \mathcal{U}\ dxdydz$$

For the KGE we have,

$$\mu(\varphi) = \frac{1}{2}\left(\frac{mc}{\hbar}\right)^2 \varphi^2 \tag{B.5}$$

then

$$\mathcal{L}(\varphi) = \frac{1}{2}\left(\frac{1}{c}\frac{\partial \varphi}{\partial t}\right)^2 - \frac{1}{2}(\nabla\varphi)^2 - \frac{1}{2}\left(\frac{mc}{\hbar}\right)^2 \varphi^2 \tag{B.6}$$

Applying the Euler-Lagrange equation[2] to $\mathcal{L}(\varphi)$,

$$\frac{\partial}{\partial t}\frac{\partial \mathcal{L}}{\partial\left(\frac{\partial \varphi}{\partial t}\right)} + \sum_{i=1}^{3}\frac{\partial}{\partial x_i}\frac{\partial \mathcal{L}}{\partial\left(\frac{\partial \varphi}{\partial x_i}\right)} - \frac{\partial \mathcal{L}}{\partial \varphi} = 0 \tag{B.7}$$

where $x_1 = x$, $x_2 = y$, and $x_3 = z$. Noticing the following,

$$\frac{\partial}{\partial x_i}\frac{\partial \mathcal{L}}{\partial\left(\frac{\partial \varphi}{\partial x_i}\right)} = \frac{\partial}{\partial x_i}\frac{\partial\left[\frac{1}{2}\left(\frac{\partial \varphi}{\partial x_i}\right)^2\right]}{\partial\left(\frac{\partial \varphi}{\partial x_i}\right)} = \frac{\partial}{\partial x_i}\frac{\partial \varphi}{\partial x_i} = \frac{\partial^2 \varphi}{\partial x_i^2}$$

we obtain the KGE,

$$\frac{1}{c^2}\frac{\partial^2 \varphi}{\partial t^2} - \nabla^2\varphi + \left(\frac{mc}{\hbar}\right)^2 \varphi = 0 \tag{B.8}$$

It turns out that the above wave equation is for a free relativistic particle of mass m.

Notice the following. Consider the plane-wave solution

$$\varphi(\boldsymbol{r},t) = \exp i(\boldsymbol{k}.\boldsymbol{r} - \omega t) = \exp\frac{i}{\hbar}(\boldsymbol{p}.\boldsymbol{r} - Et) \tag{B.9}$$

where $\boldsymbol{p}$ is the momentum and E the energy of the particle. Substitution leads to the relativistic expression for the energy of a free particle,

$$E^2 = m^2c^4 + p^2c^2$$

The ground state corresponds to the minimum of the potential: $V = 0$ at $\varphi = 0$. $V(\varphi)$ has reflection symmetry in φ in the same sense of the free energy of the ferromagnet. This particular value of φ that minimises the potential corresponds to the ground state, i.e. the *vacuum* in quantum field theory; in this case $\varphi = 0$ is the *vacuum expectation value* (vev).[3] Note that if φ is constant the kinetic energy term is zero.

B.3 GENERALISING THE LAGRANGIAN DENSITY

We can generalise the potential energy density (Eq. B.5) of the KGE. Before doing this, let us recall some basic relativistic ideas. Consider the spacetime position *fourvector*. The *contravariant* and *covariant* components are given, respectively, by

$$x^\mu = \left[x^0, x^1, x^2, x^3\right] \equiv \left[ct, x, y, z\right]$$

$$x_\mu = \left[x_0, x_1, x_2, x_3\right] \equiv \left[ct, -x, -y - z\right]$$

The contravariant and covariant expressions have the same time component, but the spatial components have opposite signs. Change to a moving coordinate system obeys the Lorentz transformation for fourvectors. The x^μ transform to a moving inertial frame according to the direct transformation, whereas the x_μ transform to a moving frame according to the inverse transformation. Any fourvector B^ν and the corresponding B_ν follow these rules. The scalar product of two fourvectors is given by

$$A^\rho B_\rho = A^0 B_0 + A^1 B_1 + A^2 B_2 + A^3 B_3$$

where we immediately observe the convention of repeated indices: each expression containing, respectively, identical covariant and contravariant indices implies a summation. Two identical contravariant or covariant indices *do not* imply summation, e.g., $A_\omega B_\omega$ and $A^\alpha B^\alpha$ do not represent summations but the products of single components.

We can also introduce the covariant and contravariant derivatives, respectively, which form fourvectors:

$$\partial_\mu \equiv \frac{\partial}{\partial x^\mu} = \left[\frac{1}{c}\frac{\partial}{\partial t} \quad \frac{\partial}{\partial x} \quad \frac{\partial}{\partial y} \quad \frac{\partial}{\partial z} \right]$$

$$\partial^\mu \equiv \frac{\partial}{\partial x_\mu} = \left[\frac{1}{c}\frac{\partial}{\partial t} \quad -\frac{\partial}{\partial x} \quad -\frac{\partial}{\partial y} \quad -\frac{\partial}{\partial z} \right]$$

The scalar product using an arbitrary scalar function $f(x)$, where the argument x of f is a shorthand notation for the position fourvector, is given by

$$\partial^\mu f \, \partial_\mu f \equiv (\partial_\mu f)^2 = \left(\frac{1}{c}\frac{\partial f}{\partial t} \right)^2 - \left(\frac{\partial f}{\partial x} \right)^2 - \left(\frac{\partial f}{\partial y} \right)^2 - \left(\frac{\partial f}{\partial z} \right)^2$$

A generalised expression for the Lagrangian density of a scalar particle represented by the field $\varphi(x)$, can be cast as,

$$\mathcal{L} = (\partial_\mu \varphi)^2 + C + A\varphi + B\varphi^2 + F\varphi^3 + G\varphi^4 + \ldots \tag{B.10}$$

In the electromagnetic field, the generation of photons occurs by exciting a mode or modes via atomic emission, for example. Thus, photons can be observable through small perturbations. Bearing this in mind, we can interpret the particle spectrum of our theory by studying the Lagrangian under small perturbations. Let us consider each of the terms in Eq. (B.10):

$(\partial_\mu \varphi)^2$: as discussed earlier, this is the kinetic energy term (recall the kinetic energy operator in 1D: $\frac{\hat{p}^2}{2m} = -\frac{\hbar^2}{2m}\frac{\partial^2}{\partial x^2}$)

C: constant potential term which is of no relevance as it does not appear in the equations of motion. It can be dropped.

$A\varphi$: this is the linear term which has no direct interpretation, and it breaks the symmetry of the Lagrangian when we replace φ by $-\varphi$. This term can be dropped.

$B\varphi^2$: this quadratic term, with $B > 0$, contains the mass of the field particle in Eq. (B.6). However, in this generalisation, B can be negative and hence is no longer the mass.

$F\varphi^3$: this is known as the three-point interaction which exhibits the same problem as the linear term in that replacing φ by $-\varphi$ breaks the symmetry of the Lagrangian. Thus, we drop it.

$G\varphi^4$: this is the quartic term that represents what is known as a four-point interaction. This is the lowest of higher order terms we can add to the theory.

B.4 EXAMPLES OF SPONTANEOUS SYMMETRY BREAKING

B.4.1 *Real scalar field* $\phi(r,t)$

Having established the terms we can add to modify the Klein-Gordon Lagrangian, we can define the model with some convenient changes in notation. Furthermore, we have the choice of signs of the potential energy density coefficients in search of new physics. Let us add on the self-interaction quartic term, so that the Lagrangian density of what we now call the *Higgs field* ϕ becomes,

$$\mathcal{L} = \mathcal{K}_\phi - \mathcal{V}(\phi) = \mathcal{K}_\phi - \left(\frac{1}{2}\mu^2\phi^2 + \frac{1}{4}\lambda\phi^4\right) \tag{B.11}$$

where K_ϕ is the kinetic energy of the field given above. If $\mu^2 > 0$ we have a particle of mass μ as before, and a plot of the potential is like Figure B.2a. The minimum of the potential that *corresponds to the vacuum* is obtained by making $\partial\mathcal{V}/\partial\phi = 0$, namely

$$0 = \phi\left(\mu^2 + \lambda\phi^2\right) \tag{B.12}$$

where $\phi = 0$ is the only solution. If we look at excitations around the vacuum, to get a particle content interpretation, by substituting in Eq. (B.5) of $\phi = 0 + \eta$, where $\eta \ll 1$, and neglecting the quartic term in the Lagrangian density, we obtain the KGE Lagrangian density with the same mass and symmetry, where $\phi = 0$ is the only solution.

$$\begin{aligned}
\mathcal{L}(\eta) &= \frac{1}{2}\left(\frac{1}{c}\frac{\partial\eta}{\partial t}\right)^2 - \frac{1}{2}(\nabla\eta)^2 - \frac{1}{2}\left(\frac{mc}{\hbar}\right)^2\eta^2 \\
\mathcal{L}(\eta) &= \mathcal{L}(-\eta) \\
V(\eta) &= V(-\eta) = \frac{1}{2}\left(\frac{mc}{\hbar}\right)^2\eta^2
\end{aligned} \tag{B.13}$$

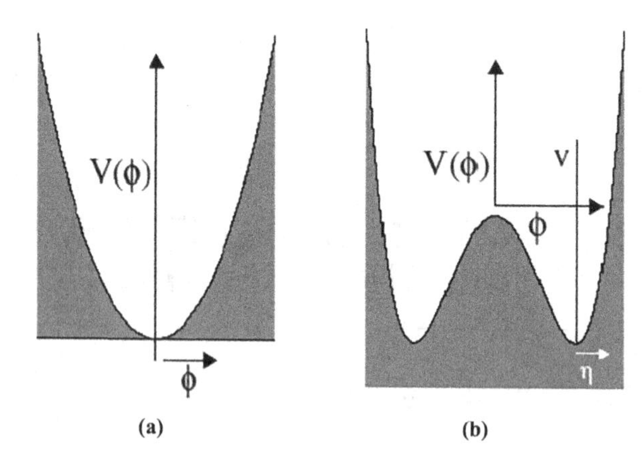

FIGURE B.2 Potential energy density $\mathcal{V}(\phi)=\frac{1}{2}\mu^2\phi^2+\frac{1}{4}\lambda\phi^4$. (a) $\mu^2>0$. In this case we get a parabolic shape with a minimum at $\phi=0$ and μ is the mass. The reflection symmetry $\mathcal{V}(\phi)=\mathcal{V}(-\phi)$ is preserved. (b) $\mu^2<0$. Here μ is no longer the mass. The potential density has two minima: $\mathcal{V}(\phi)=\mathcal{V}(-\phi)$. The quantity η is an excitation around the minimum (vacuum). It turns out that $\mathcal{V}(\eta)\neq\mathcal{V}(-\eta)$ and thus the symmetry is spontaneously broken (Courtesy: Ivo van Lupen).

When $\mu^2<0$ we get the curve $V(\phi)$ of Figure B.2b. There are three solutions to Eq. (B.9). One is $\phi=0$ which gives a maximum, and the other two are $\phi_\pm=\pm\sqrt{-\frac{\mu^2}{\lambda}}$. This shows us that the vev is not zero and can acquire two possible values, $\pm v$, where $v\equiv+\sqrt{-\frac{\mu^2}{\lambda}}$. As opposed to $\mu^2>0$, $\phi=0$ is no longer a vacuum state as it corresponds to a maximum of the potential. Substitute $\phi=v+\eta$ in Eq. (B.10), namely,

$$\mathcal{L}(\varphi)=\frac{1}{2}\left(\frac{1}{c}\frac{\partial\phi}{\partial t}\right)^2-\frac{1}{2}(\nabla\phi)^2+\frac{1}{2}\mu^2\phi^2+\frac{1}{4}\lambda\phi^4 \tag{B.14}$$

and we get

$$\mathcal{L}(\eta)=\frac{1}{2}\left(\frac{1}{c}\frac{\partial\eta}{\partial t}\right)^2-\frac{1}{2}(\nabla\eta)^2-\frac{1}{2}2\lambda v^2\eta^2-\lambda v\eta^3-\frac{1}{4}\lambda\eta^4$$

Notice the potential density,

$$\mathcal{V}(\eta)=\frac{1}{2}2\lambda v^2\eta^2+\lambda v\eta^3+\frac{1}{4}\lambda\eta^4\neq\mathcal{V}(-\eta)$$

showing the loss of reflection symmetry, and thus the Lagrangian, since $\mathcal{L}(\eta)\neq\mathcal{L}(-\eta)$.

Comparison with the KGE tells us about the appearance of a particle, there was none before, with a mass m_η proportional to the vev of the field v:

$$m_\eta=\sqrt{2\lambda}v \tag{B.15}$$

Assuming $\mu^2<0$, the particle spectrum revealed by the fluctuations (excitations) around the minimum is a massive scalar particle with self-interactions. Notice that the Lagrangian density retains the original reflection symmetry $\mathcal{L}(\phi)=\mathcal{L}(-\phi)$ but the vacuum is not symmetric in the field: this is *spontaneous symmetry breaking* analogous to the appearance of a net magnetisation in the ferromagnet.

This relatively simple example tells us how particles acquire mass even when the original Lagrangian was indicating that there was no mass present. Notice in Eq. (B.9) that the mass of the particle also depends on the fourth-order self-interaction, namely the field interacting with itself to provide the mass of the particles. Basically, it is energy converted into mass. However, it misses an important complement, namely *gauge symmetry*. This latter is part and parcel of the *Higgs mechanism*, which together with spontaneous symmetry breaking provides mass to gauge bosons as explained by the electroweak theory. Furthermore, it provides mass to fermions, i.e., quarks and leptons, via the Yukawa coupling. Protons and neutrons are fermions but their mass stems not only from quarks but also by the internal energies from quark interaction with gluons. A proper mathematical treatment is left for Appendix C, which gives an introduction to the electroweak theory.

B.4.2 Complex scalar field $\phi(\boldsymbol{r},t)$

Suppose we consider a complex Higgs field with a "Mexican hat" potential density (Figure B.3),

$$\mathcal{V}(\phi,\phi^*)=\frac{1}{2}\mu^2\phi^*\phi+\frac{1}{4}\lambda(\phi^*\phi)^2 \quad \text{with } \mu^2<0,\lambda>0$$

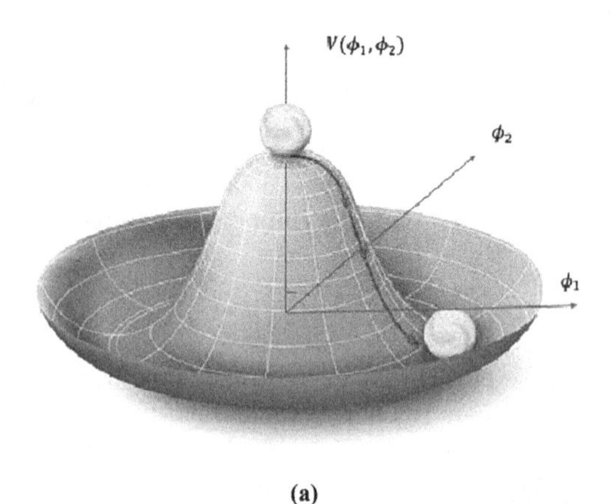

(a)

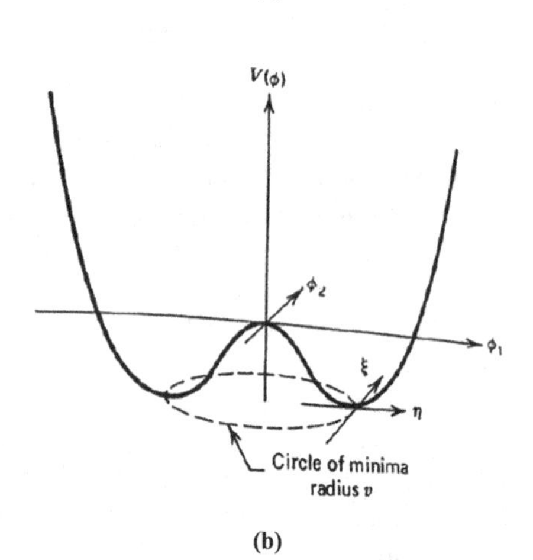

(b)

FIGURE B.3 "Mexican hat" Higgs potential density for a complex scalar field $\phi_1 + i\phi_2$exhibiting spontaneous symmetry breaking. This is a revolution surface obtained by rotating Figure B.2 (b) around the vertical axis. Its minima lie on a circumference of radius v, the vev, on a plane parallel to the complex $\phi_1 - \phi_2$ plane.

Thus, the Lagrangian density

$$\mathcal{L} = \frac{1}{2}\left(\partial_\mu \phi\right)^* \left(\partial^\mu \phi\right) - \frac{\mu^2}{2}\phi^*\phi - \frac{\lambda}{4}\left(\phi^*\phi\right)^2 \tag{B.16}$$

has a global symmetry, whereby $\phi \rightarrow e^{i\theta}\phi$ will not be changing the Lagrangian.[4] Then following the same steps as before and expressing the complex field in terms of its components $\phi = \phi_1 + i\phi_2$ we get the following equation for the minima:

$$\phi_1^2 + \phi_2^2 = v^2$$

i.e., this is a circumference of radius v in the complex $\phi_1 - \phi_2$ plane. Due to the cylindrical symmetry of the potential, we can look at excitations anywhere on the circumference to reveal the particle spectrum. In particular we choose small perturbations η and ξ as follows:

$$\phi_1 = v + \eta \qquad \phi_2 = i\xi$$

Following substitution in the Lagrangian density we obtain,

$$\mathcal{L}(\eta,\xi) = \frac{1}{2}\left[\left(\frac{\partial\eta}{\partial t}\right)^2 - (\nabla\eta)^2 - 2\lambda v^2\eta^2\right]$$

$$+\frac{1}{2}\left(\frac{\partial\xi}{\partial t}\right)^2 - (\nabla\xi)^2 + \text{other irrelevant terms}$$

The terms in brackets refer to the Higgs boson where the mass term is clearly identifiable by the vev, namely $-2\lambda v^2\eta^2$. Regarding ξ, there is no mass term as there is not something like $m_\xi\xi^2$. Yet we have kinetic energy $\frac{1}{2}\left(\frac{\partial\xi}{\partial t}\right)^2 - (\nabla\xi)^2$. We have what is known as a *massless Goldstone boson.*

Here it is more intuitive as in the ξ direction there is no potential gradient and a massless mode can be sustained.

NOTES

1 In a second-order phase transition, the change of phase occurs gradually as opposed to a first-order phase transition where an amount of heat is delivered or extracted, known as *latent heat*, while the temperature remains constant.

2 Any book on mechanics explains this equation and how to apply it. Check, e.g., *Mechanics by Goldstein.*

3 The word “expectation” is used because of quantum fluctuations of the field around the classical value. The vev should be written $\langle \varphi \rangle$.

4 For $\theta = \pi$ this includes the reflection we had before.

Appendix C

The Standard Model of Particle Physics III

How gauge bosons acquire mass and electroweak unification

The introduction of gauge fields in quantum field theory has been striking, as we discuss in this appendix. Gauge is a particular choice of name which in fact means measurement. Gauge fields appeared first in the context of electromagnetism. Without further ado, we show how this works in Maxwell's equations.

C.1 GAUGE SYMMETRY

Electromagnetism provides a form of symmetry which manifests itself via gauge transformations which were key to the development of particle physics. What are they about?

We introduce them through Maxwell's equations, which in the Heaviside-Lorentz units read,

$$\nabla \cdot \boldsymbol{E} = \rho$$

$$\nabla \times \boldsymbol{E} = -\frac{\partial \boldsymbol{B}}{\partial t} \tag{C.1}$$

$$\nabla \cdot \boldsymbol{B} = 0 \tag{C.2}$$

$$\nabla \times \boldsymbol{B} = \boldsymbol{J} + \frac{\partial \boldsymbol{E}}{\partial t}$$

where

- $\boldsymbol{E}$: electric field
- $\boldsymbol{B}$: magnetic field
- $\boldsymbol{J}$: current density
- ρ: charge density

Let us recall some identities. If $\boldsymbol{V}(\boldsymbol{r})$ is a vector field and $f(\boldsymbol{r})$ a well-behaved function, then the following holds:

$$\nabla \cdot \nabla \times \boldsymbol{V} \equiv 0 \tag{C.3}$$

$$\nabla \times \nabla f \equiv 0 \tag{C.4}$$

Using Eqs. (C.3) and (C.4) in Eqs. (C.2) and (C.1), respectively, we can define a scalar potential ϕ and a vector potential $\boldsymbol{A}$ such that,

$$\boldsymbol{B} = \nabla \times \boldsymbol{A} \tag{C.5}$$

$$\boldsymbol{E} = -\nabla \phi - \frac{\partial \boldsymbol{A}}{\partial t} \tag{C.6}$$

The conservation of charge is expressed through the continuity equation:

$$\frac{\partial \rho}{\partial t} + \nabla \cdot \boldsymbol{J} = 0$$

As the charges are conserved in all frames, we can define a current density fourvector

$$j^{\mu} = (\rho, \boldsymbol{J})$$

where the time component is the charge density, and the spatial component is the current density vector. Now we can write the continuity equation as follows:

$$\partial_{\mu} j^{\mu} = 0$$

which looks like relativistic scalar product and therefore invariant under a change of reference inertial frame.

C.2 LOCAL GAUGE INVARIANCE

By local gauge variance, we understand multiplying the field by a complex number and obtain invariance in the Lagrangian. This is the $U(1)$ symmetry discussed in Appendix A. Just take the Lagrangian of a complex field as in Eq. (B.9)

$$\mathcal{L} = \frac{1}{2}\left(\partial_\mu \phi\right)^* \left(\partial^\mu \phi\right) - \frac{\mu^2}{2}\phi^*\phi - \frac{\lambda}{4}\left(\phi^*\phi\right)^2 \tag{C.7}$$

If we now replace $\phi \to e^{i\theta}\phi$ with θ being a constant, it is trivial to see the cancellation of phases in every term, i.e., we have a global $U(1)$ symmetry (Appendix A).

However, if we substitute θ by an arbitrary function $\Theta(x)$, where x is a symbol for the four spacetime variables $(ct, \boldsymbol{r})$, and replace $\phi(x)$ by

$$\phi'^{(x)} = e^{ig\Theta(x)}\phi(x) \tag{C.8}$$

this change does not affect the potential energy. On the other hand, the kinetic energy has derivatives which will act on the exponential. In other words, the former global symmetry is lost, and we do not even have a local symmetry.

The way of going about this problem is by introducing a *covariant derivative*. This is defined as follows:

$$D_\mu \equiv \partial_\mu + igB_\mu \tag{C.9}$$

where B_μ is an undefined gauge field in the same manner as we did when we discussed Maxwell's equations. The resulting Lagrangian is

$$\mathcal{L} = \frac{1}{2}\left(D_\mu \phi\right)^* \left(D^\mu \phi\right) - \frac{\mu^2}{2}\phi^*\phi - \frac{\lambda}{4}\left(\phi^*\phi\right)^2 \tag{C.10}$$

If we transform the field as $\phi'(x) = e^{ig\Theta(x)}\phi(x)$ then we have to establish a transformation rule for the gauge field as well:

$$B'_\mu(x) = B_\mu(x) - \partial_\mu\Theta(x) \tag{C.11}$$

This means the following. Take

$$\left(D_\mu\phi(x)\right)^* = \left(\partial_\mu - igB_\mu\right)\phi^*(x)$$

Now we replace $\phi(x)^*$ by $e^{-ig\Theta(x)}\phi^*(x)$ and simultaneously $B_\mu \to B'_\mu = B_\mu - \partial_\mu\Theta(x)$:

$$\begin{aligned}\left[\partial_\mu - ig(B_\mu - \partial_\mu\Theta\right]e^{-ig\Theta}\phi^* &= -ig\,\partial_\mu\Theta\; e^{-ig\Theta}\phi^* \\ &\quad + e^{-ig\Theta}\,\partial_\mu\phi^* - igB_\mu\; e^{-ig\Theta}\phi^* + ig\,\partial_\mu\Theta\; e^{-ig\Theta}\phi^* \\ &= e^{-ig\Theta}\,\partial_\mu\phi^* - igB_\mu\; e^{-ig\Theta}\phi^* \\ &= e^{-ig\Theta}\left(\partial_\mu\phi^* - igB_\mu\phi^*\right)\end{aligned} \tag{C.12}$$

In a similar fashion

$$\left[\partial^\mu + ig(B^\mu - \partial^\mu\Theta\right]e^{ig\Theta}\phi^\mu = e^{ig\Theta}\left(\partial^\mu\phi + igB^\mu\phi\right) \tag{C.13}$$

The kinetic energy is the product of Eqs. (C.12) and (C.13), as a result of which, the phases cancel each other out and we recover the original kinetic energy of the Lagrangian.

C.3 THE HIGGS MECHANISM

We have already discussed spontaneous symmetry breaking. The Higgs mechanism is about symmetry breaking of a complex scalar field with a potential,

$$V(\phi) = \frac{\mu^2}{2}\phi^*\phi + \frac{\lambda}{4}\left(\phi^*\phi\right)^2$$

and this potential is set against local gauge symmetry. As an example, we choose a $U(1)$ local gauge symmetry. However, it is not the only type of gauge symmetry. For the electroweak discussion we have to use an SU(2) symmetry, which is a rotational gauge symmetry (Appendix A).

The establishment of a local gauge symmetry introduced a new field, B_μ. In principle this is a massless field, which we require. The reason for the latter is that if it had a mass m_B there would be a kinetic energy term like $\frac{1}{2}m_B B_\mu B^\mu$ which would break the gauge symmetry. This latter can be checked as follows:

$$\frac{1}{2}m_B B_\mu B^\mu \to \frac{1}{2}m_B B'_\mu B'^\mu = \frac{1}{2}m_B\left(B_\mu - \partial_\mu \Theta\right)\left(B^\mu - \partial^\mu \Theta\right) \neq \frac{1}{2}m_B B_\mu B^\mu$$

Having assumed that we are dealing with a massless field and in anticipation of discussing electroweak theory, this massless particle will be the photon. As such we can write the kinetic energy in the Lagrangian as (Appendix B)

$$-\frac{1}{4}F_{\mu\nu}F^{\mu\nu}$$

where

$$F^{\mu\nu} \equiv \partial^\mu B^\nu - \partial^\nu B^\mu$$

Thus, the Lagrangian,

$$L = -\frac{1}{4}F_{\mu\nu}F^{\mu\nu} + \frac{1}{2}\left(D_\mu\phi\right)^*\left(D^\mu\phi\right) - \frac{\mu^2}{2}\phi^*\phi - \frac{\lambda}{4}\left(\phi^*\phi\right)^2 \tag{C.14}$$

noting that

$$\left(D_\mu\phi\right)^*\left(D^\mu\phi\right) = \left(\partial_\mu\phi\right)^*\left(\partial^\mu\phi\right) - igB_\mu\phi^*\partial_\mu\phi + ig\left(\partial^\mu\phi^*\right)B^\mu\phi + g^2 B_\mu B^\mu\phi^*\phi$$

Once again taking $\mu < 0$ the vacuum state is degenerate along a circumference parallel to the complex plane whose radius is the vacuum expectation value v which we already defined (Appendix B). Due to the cylindrical symmetry, all points of the circumference are equivalent. However, choosing arbitrarily some point on the circumference breaks the symmetry. We choose

$$\phi(x) = \phi_1(x) + i\phi_2(x) = v$$

To get the excitations we expand around this value $(v, 0)$:

$$\phi = \frac{1}{\sqrt{2}}\left(v + \eta(x) + i0 + i\xi(x)\right) \tag{C.15}$$

Substituting Eq. (C.15) into Eq. (C.14), the Lagrangian reads,

$$\mathcal{L}=\frac{1}{2}\partial_{\mu}\eta\,\partial^{\mu}\eta-\lambda v^{2}\eta^{2}+\frac{1}{2}\partial_{\mu}\xi\,\partial^{\mu}\xi-\frac{1}{4}F_{\mu\nu}F^{\mu\nu}+\frac{1}{2}gv^{2}B_{\mu}B^{\mu}-V_{i}+gvB_{\mu}\,\partial^{\mu}\xi \tag{C.16}$$

Let us recall the Lagrangian for the Klein-Gordon equation:

$$\mathcal{L}_{\mathrm{KGE}}=\frac{1}{2}\partial^{\nu}\varphi\,\partial_{\nu}\varphi-\frac{1}{2}m^{2}\varphi^{2}$$

It tells us how to identify whether the field has a mass associated with it. Thus, we have to look for terms which are identical in structure to that of the Klein-Gordon Lagrangian, namely

$$\frac{1}{2}m^{2}\varphi^{2}$$

On inspection of Eq. (C.16) we come to the following conclusions:

$\frac{1}{2}\partial_{\mu}\eta\,\partial^{\mu}\eta-\lambda v^{2}\eta^{2}$: massive particle η with mass $m_{\eta}=\sqrt{2\lambda v^{2}}$

$\frac{1}{2}\partial_{\mu}\xi\,\partial^{\mu}\xi$: massless particle ξ (Goldstone boson)

$-\frac{1}{4}F_{\mu\nu}F^{\mu\nu}+\frac{1}{2}gv^{2}B_{\mu}B^{\mu}$: massive gauge field B with mass $m_{B}=gv$

V_i : self-interactions

$gvB_{\mu}\,\partial^{\mu}\xi$: coupling between the gauge field and the Goldstone boson.

The symmetry breaking causes the appearance of a massive scalar field η and the massless Goldstone boson ξ. Quite remarkably, the previously massless gauge field B acquires mass, which points at the way of how to give mass to gauge bosons.

C.4 THE HIGGS FIELD

For reasons that are beyond the scope of this book, the fields appearing in Eq. (C.16) are not physical fields as there is one more degree of freedom than the original Lagrangian. We can, however, eliminate the Goldstone degree of freedom by a suitable gauge transformation.

Notice the following. From Eq. (C.16) we can take three terms and express them in a more compact way

$$\frac{1}{2}\partial_\mu\xi\,\partial^\mu\xi+\frac{1}{2}gv^2B_\mu B^\mu+gvB_\mu\,\partial^\mu\xi=\frac{1}{2}g^2v^2\left(B_\mu+\frac{1}{gv}\partial_\mu\xi\right)^2 \tag{C.17}$$

Note that

$$\frac{1}{2}g^2v^2\left(B_\mu+\frac{1}{gv}\partial_\mu\xi\right)^2=\frac{1}{2}g^2v^2\left(B_\mu+\frac{1}{gv}\partial_\mu\xi\right)\left(B^\mu+\frac{1}{gv}\partial^\mu\xi\right) \tag{C.18}$$

The right-hand side of Eq. (C18) suggests that we can do a new gauge transformation by defining

$$B'_\mu=B_\mu+\frac{1}{gv}\partial_\mu\xi$$

and the Lagrangian becomes

$$\mathcal{L}=\frac{1}{2}\partial_\mu\eta\,\partial^\mu\eta-\lambda v^2\eta^2-\frac{1}{4}F_{\mu\nu}F^{\mu\nu}+\frac{1}{2}gv^2B'_\mu B'^\mu-V_i$$

And we get a massive η and a massive gauge field. The last gauge transformation does not affect the underlying physics. In fact, we will see that this gauge transformation will elucidate new physics.

In the original gauge transformation, we had $e^{ig\Theta(x)}\phi(x)$ which now we replace by

$$\Theta(x)=-\xi(x)/gv\,.$$

Thus,

$$\phi'(x)=e^{\frac{-i\xi(x)}{v}}\phi(x)$$

The transformations we made do not affect the vacuum state. Recall that we have obtained the vacuum state by working purely with the potential and the potential is exactly the same. Hence, after the symmetry breaking caused by the potential, we can once again expand the physical field around the vacuum state:

$$\phi(x)=\frac{1}{\sqrt{2}}(v+\eta(x)+i\xi(x))$$

Notice the following:

$$(v+\eta(x))e^{\frac{i\xi}{v}} = (v+\eta(x))(1+i\xi/v + \text{2nd order terms}\ldots)$$

Then to first order

$$= (v+\eta(x)+i\xi(x))$$

and the physical field around the vacuum state reads,

$$\phi(x) = \frac{1}{\sqrt{2}}(v+\eta(x))e^{\frac{i\xi}{v}}$$

And now we perform the new gauge transformation

$$\phi'(x) = e^{-\frac{i\xi}{v}}\phi(x) = e^{\frac{-i\xi}{v}}\frac{1}{\sqrt{2}}(v+\eta(x))e^{\frac{i\xi}{v}}$$

$$\phi'(x) = \frac{1}{\sqrt{2}}(v+\eta(x)) \tag{C.19}$$

Equation (C.19) shows us that we managed to convert the complex scalar field to being completely real by eliminating the Goldstone boson through a unitary transformation (Appendix A). We can identify $\eta(x)$ with the Higgs field $h(x)$, namely the physical field resulting from the unitary transformation. Substituting Eq. (C.19) in the original Lagrangian, namely

$$L = -\frac{1}{4}F_{\mu\nu}F^{\mu\nu} + \frac{1}{2}(D_\mu\phi)^*(D^\mu\phi) - \frac{\mu^2}{2}\phi^*\phi - \frac{\lambda}{4}(\phi^*\phi)^2 \tag{C.20}$$

we get,

$$\mathcal{L} = \frac{1}{2}(\partial_\mu h\,\partial^\mu h - \lambda v^2 h^2) - \frac{1}{4}F_{\mu\nu}F^{\mu\nu} \tag{C.21}$$

C.5 ELECTROWEAK THEORY: MASSES OF THE $W^{\pm}$, Z^0, AND ZERO MASS OF THE PHOTON

We will now place the Higgs mechanism in the context of the Standard Model. In particular we wish to establish that through the Higgs mechanism we can confirm the experimental findings that the gauge bosons of the weak force, $W^{\pm}$ and Z^0, are indeed massive. Furthermore, we show that the photon appears without a mass, thus setting the foundations for the electroweak unification. The gauge symmetry will then consist of $U(1)$ for the electromagnetic sector, and $SU(2)$ for the weak force sector. This symmetry is expressed as $U(1) \times SU(2)$ local symmetry and this is where the Higgs mechanism is embedded. This is the foundation of the Weinberg-Salam-Glashow model for which they got the Nobel Prize.

The minimal Higgs model consists of two complex fields forming a 2×1 vector. The following is the standard notation:

$$\phi = \begin{pmatrix} \phi^+ \\ \phi^0 \end{pmatrix} = \frac{1}{\sqrt{2}} \begin{pmatrix} \phi_1 + i\phi_2 \\ \phi_3 + i\phi_4 \end{pmatrix} \tag{C.22}$$

This is known as the *weak isospin doublet* in which the upper and lower components differ by a unit of charge. Thus, the Lagrangian reads,

$$\mathcal{L}_W = \frac{1}{2}\partial_\nu \phi^\dagger \partial^\nu \phi - \frac{1}{2}\mu^2 \phi^\dagger \phi - \frac{1}{4}\lambda \left(\phi^\dagger \phi\right)^2 \tag{C.23}$$

$\phi^\dagger$ is the Hermitian conjugate of ϕ

$$\phi^\dagger = \frac{1}{\sqrt{2}} \begin{pmatrix} \phi_1 - i\phi_2 & \phi_3 - i\phi_4 \end{pmatrix}$$

with

$$\phi^\dagger \phi = \frac{1}{2} \begin{pmatrix} \phi_1 - i\phi_2 & \phi_3 - i\phi_4 \end{pmatrix} \begin{pmatrix} \phi_1 + i\phi_2 \\ \phi_3 + i\phi_4 \end{pmatrix} = \frac{1}{2}\left(\phi_1^2 + \phi_2^2 + \phi_3^2 + \phi_4^2\right)$$

For $\mu < 0$ the potential has a set of infinite minima on a circumference of radius v so that

$$\frac{1}{2}\left(\phi_1^2 + \phi_2^2 + \phi_3^2 + \phi_4^2\right) = \frac{v^2}{2} = \frac{-\mu^2}{2\lambda}$$

After symmetry breaking, the photon must remain massless. This is the opposite of what we were getting so far. Only ϕ_0 is neutral, so the minimum of the potential must correspond to the expectation value of ϕ_0 :

$$\langle|\phi|\rangle = \frac{1}{\sqrt{2}} \begin{pmatrix} 0 \\ v \end{pmatrix}$$

Thus, repeating what we did earlier to do away with a Goldstone boson, we can write

$$\phi(x) = \frac{1}{\sqrt{2}}\begin{pmatrix} 0 \\ v + h(x) \end{pmatrix} \tag{C.24}$$
$$h(x): \text{Higgs field}$$

We are interested in getting the masses of the gauge fields which ultimately will give us the masses of the electroweak gauge bosons. We then proceed by introducing suitable covariant derivatives which will take into account the three independent generators of the $SU(2)$ symmetry. These three independent generators will account for the three weak force particles, but in addition we would have an extra gauge field to account for the photon. So altogether we will have four fields. Hence,

$$D_\mu = \partial_\mu + ig_W \boldsymbol{T} \cdot \boldsymbol{W}_\mu + ig' \frac{Y}{2} B_\mu \tag{C.25}$$

$$\boldsymbol{T} = \frac{\boldsymbol{\sigma}}{2} = \frac{1}{2}(\sigma_1, \sigma_2, \sigma_3)$$

where $\sigma_1, \sigma_2, \sigma_3$ are the Pauli matrices and the generators of the $SU(2)$ group (Appendix A).

$$\sigma_1 = \begin{pmatrix} 0 & 1 \\ 1 & 0 \end{pmatrix} \sigma_2 = \begin{pmatrix} 0 & i \\ -i & 0 \end{pmatrix} \sigma_3 = \begin{pmatrix} 1 & 0 \\ 0 & -1 \end{pmatrix}$$

The gauge fields are given by

$$\boldsymbol{W} = \left(W_\mu^{(1)}, W_\mu^{(2)}, W_\mu^{(3)}\right) \text{ and } B_\mu$$

$$Y = 1, \text{ weak hypercharge}$$

The covariant derivative acting on the Higgs doublet reads,

$$D_\mu \phi = \frac{1}{2}\left[2I\,\partial_\mu + ig_W \boldsymbol{\sigma} \cdot \boldsymbol{W}_\mu + ig' B_\mu\right]\phi$$

Notice that D_μ is a 2×2 matrix just like the Pauli matrices. In the expression of the kinetic energy, we obtain a scalar product given by $\left(D_\mu \phi\right)^\dagger \left(D_\mu \phi\right)$.

From this scalar product, we obtain a sum of terms. From these, we pick out those that can provide a mass according to what we established in connection with the Klein-Gordon Lagrangian. These mass terms are quadratic in the fields, and they are the following:

$$\frac{1}{8}g_W^2v^2\left(W_\mu^{(1)}W^{(1)\mu}+W_\mu^{(2)}W^{(2)\mu}\right)+\frac{1}{8}v^2\left(g_WW_\mu^{(3)}-g'B_\mu\right)\left(g_WW^{(3)\mu}-g'B^\mu\right) \tag{C.26}$$

The first bracket of Eq. (C.26) clearly exhibits the masses for the $W^{(1)}$ and $W^{(2)}$ being identical:

$$\frac{1}{2}m_W^2W_\mu^{(1)}W^{(1)\mu} \quad \text{and} \quad \frac{1}{2}m_W^2W_\mu^{(2)}W^{(2)\mu}$$

$$m_W=\frac{g_Wv}{2}$$

Thus, $W^{(3)}$ will correspond to the weak gauge boson Z^0 and B_μ to the photon. We can rewrite the product of brackets as follows:

$$\frac{1}{8}v^2\left(g_WW_\mu^{(3)}-g'B_\mu\right)\left(g_WW^{(3)\mu}-g'B^\mu\right)=\frac{v^2}{8}\left(W_\mu^{(3)}\quad B_\mu\right)\mathbf{M}\begin{pmatrix}W^{(3)\mu}\\B^\mu\end{pmatrix}$$

where $\mathbf{M}$ is a non-diagonal mass matrix,

$$\mathbf{M}\equiv\begin{pmatrix}g_W^2 & -g_Wg'\\-g_Wg' & g'^2\end{pmatrix}$$

Note that the off-diagonal elements couple the $W^{(3)}$ and B fields, allowing them to mix. The physical boson fields correspond to a basis in which the matrix is diagonal, free of mixing. Thus, the masses of the physical gauge bosons are given by the eigenvalues of $\mathbf{M}$ obtained from the standard characteristic equation $\det(\mathbf{M}-\lambda I)=0$. Performing this leads to the eigenvalues

$$\lambda_1=0 \quad \lambda_2=g_W^2+g'^2$$

These are the diagonal elements of the diagonalised matrix

$$\begin{bmatrix} m_A^2 & 0 \\ 0 & m_Z^2 \end{bmatrix}$$

with $m_A = 0$ (photon) and $m_Z = \frac{1}{2}v\sqrt{g_W^2 + g'^2}$ (Z gauge boson).

These truly beautiful results stem from the electroweak unification of Glashow-Weinberg-Salam. However, there are four parameters unknown: the two couplings and the Higgs potential parameters μ and λ. So how can we progress quantitatively? It is down to measurements in the Large Hadron Collider (LHC) in Switzerland. From measurements it is possible to obtain m_W and g_W and use these measurements in the expression

$$m_W = \frac{1}{2}g_W v$$

to obtain $v = 246$ GeV. The mass of the Higgs boson is given by $m_H^2 = 2\lambda v^2$. The parameter λ is obtained from measurements of the Higgs boson mass at LHC.

Index

Note: **Bold** page numbers refer to tables; *italic* page numbers refer to figures and page numbers followed by "n" denote endnotes.